Bernhard Siegfried Albinus

The Anatomy of Painting

Or, a short and easy introduction to anatomy : being a new edition, on a smaller scale, of six tables of Albinus, with their linear figures

Bernhard Siegfried Albinus

The Anatomy of Painting
Or, a short and easy introduction to anatomy : being a new edition, on a smaller scale, of six tables of Albinus, with their linear figures

ISBN/EAN: 9783337606466

Printed in Europe, USA, Canada, Australia, Japan

Cover: Foto ©berggeist007 / pixelio.de

More available books at **www.hansebooks.com**

THE

ANATOMY OF PAINTING:

OR A

SHORT AND EASY INTRODUCTION TO ANATOMY:

BEING A NEW EDITION, ON A SMALLER SCALE,
OF SIX TABLES OF ALBINUS, WITH THEIR LINEAR FIGURES:
ALSO, A NEW TRANSLATION OF ALBINUS'S HISTORY OF THAT WORK,
AND OF HIS INDEX TO THE SIX TABLES:

TO WHICH ARE ADDED THE ANATOMY OF CELSUS, WITH NOTES,
AND THE PHYSIOLOGY OF CICERO:

WITH AN INTRODUCTION, GIVING A SHORT VIEW
OF PICTURESQUE ANATOMY.

By JOHN BRISBANE, M. D.

SIC EGO NUNC, QUONIAM HÆC RATIO PLERUMQUE VIDETUR
TRISTIOR ESSE, QUIBUS NON EST TRACTATA, RETROQUE
VOLGUS ABHORRET AB HAC: VOLUI TIBI SUAVILOQUENTI
CARMINE PIERIO RATIONEM EXPONERE NOSTRAM,
ET QUASI MUSEO DULCI CONTINGERE MELLE,
SI TIBI FORTE ANIMUM TALI RATIONE TENERE
VERSIBUS IN NOSTRIS POSSEM, DUM PERSPICIS OMNEM
NATURAM RERUM, QUA CONSTET COMPTA FIGURA.

LUCRET.

LONDON:
PRINTED BY GEORGE SCOTT,
AND SOLD BY T. CADELL BOOKSELLER, SUCCESSOR TO MR. MILLAR, IN THE STRAND.
MDCCLXIX.

THE GENERAL PREFACE.

IT is doing ſervice to mankind, to extend the knowledge of any uſeful art, and to render it more eaſy and delightful. Many would incline to have ſome knowledge of the ſtructure of the human frame, who cannot bear dead bodies, and actual diſſection, and who cannot reliſh the common dry and tedious method of treating anatomy. Diſſections of the human body, in ſchools of anatomy, are by far too frequent, and often to very little purpoſe; on the contrary, it would be an eaſy matter to teach the moſt uſeful part of anatomy, by models and figures alone, properly explained; ſurely ſo far as to ſatisfy every one, except thoſe who ſtudied it as a preparation for the practice of phyſic and ſurgery; and even in that caſe, anatomy might be taught in a much ſhorter and more agreeable manner than is commonly done. Many writers upon arts and ſciences, and particularly on anatomy, ſeem to have no juſt idea of the nature and utility of figures, and moſt anatomiſts confine that art, almoſt entirely, to the purpoſes of phyſic and ſurgery; whereas it is neceſſary in a particular manner, to thoſe who ſtudy and practiſe the arts of deſign, and ought to be taught and wrote upon with that view, by men ſkilful in theſe arts. For tho' phyſicians and ſurgeons have, for a long time, in a manner engroſſed the whole buſineſs of teaching anatomy; yet painters, ſtatuaries, and engravers, ſhould aſſert their rights, and teach, and write upon this ſcience, in a pictureſque manner, ſuited to their own art; in which it is as immediately and eſſentially uſeful, as even in phyſic or ſurgery; for no one can poſſibly treat of anatomy for the uſe of painters, ſo as to ſatisfy and be agreeable to true judges and lovers of the arts of deſign, unleſs he himſelf is ſkilled in theſe arts, and in the true application of anatomy to painting. Let none however imagine, becauſe from experience of their great utility, I ſo much recommend models and figures, as a good introduction to anatomy, that I do not see the use and necessity of actual dissection, to those who would be complete masters of this science; and especially in the art of medicine, in order to discover the seats and causes of diseases. I want only to control the manifest abuses of dissection, as the useful part of the structure of the human body is now fully ascertained, and to show, that good figures are on many occasions much more proper, useful, and instructive, as an introduction to anatomy, tho' little understood or recommended by most teachers of that art, who are seldom judges of good or bad figures, tho' some have given pompous ones, wherein they have been totally misled by artists unskilful in these matters. The greatest men of antiquity, looked upon it as part of a liberal education, to have some idea of every important science, and among the rest of anatomy; accordingly we find a short but elegant system of the animal œconomy, in Cicero's second book of the Nature of the Gods; and the like of anatomy in the works of Celsus, comprehended in the compass of a few pages, in such a manner, as must give the greatest pleasure to every lover of that art. How different from the tedious manner of most anatomical authors? For it is to be lamented, that not only the common herd, but even some great anatomists, carried away by the irresistible torrent of custom, have often descended to too great minuteness; but what is tolerable, and sometimes useful and agreeable, in these first rate men, in a vulgar anatomist, who has minuteness alone to recommend him, is in the highest degree tiresome and disgustful; as he is not capable to adorn his subject with any thing either useful or elegant. As to anatomical figures, tho' at present too much neglected, they were not only used by the ancients, and early introduced by the great restorers of modern anatomy, but are pursued and improved to this day, with great judgment and skill, by the most learned and elegant anatomists; and their construction and use ought in a particular manner to be understood, by those who teach or who study anatomy with a view to the arts of design, tho' they are most worthy of the study and attention of anatomists of every kind. Accordingly

Vesalius

Vesalius, physician to the emperor Charles V. the great restorer of anatomy and surgery among the moderns, in his immortal work on the fabrick of the human body, gave admirable figures of the bones and muscles, especially the external orders: he lived at a time, when painting flourished in Italy to a very high degree; and many of his figures, not only give the greatest delight and instruction to an anatomist, but also to a painter. His great rival Eustachius pursued a different plan, more accurate in anatomical truth, diligent in the highest degree, learned and ingenious; but without the noble fire, and manly freedom of Vesalius. The figures of Eustachius are the most valuable part of his remains, and, tho' as to the painter's art simple and unadorned, are most clear and instructive to an anatomist, and executed with great judgment; so that even in our days, they are held in the highest esteem; and indeed neither the particulars they contain (and they may be said to contain almost an entire system of the anatomy that was then known) nor the skilful manner in which they were constructed, were understood, till pointed out and explained by the great Albinus; who, together with the utmost judgment and accuracy, has added true elegance to every part of anatomy, but in a particular manner to anatomical figures, and has even in a manner accomodated them to the art of painting. And if I may judge from the labour it cost me, to follow his footsteps, in an exact copy of some part of his works, the labours he himself has undergone, must have exhausted the utmost extent of human patience; especially in one so eminent for elegance of genius, a character generally so impatient of tedious and laborious pursuits.

The six tables of Albinus, which are now published in a smaller form, tho' they may serve as an introduction to anatomy, and as an ornament to hang up in the studies of such as love that science, are chiefly intended for the use of those, who pursue the arts of design, in order to awaken their attention to this part of their profession, and as a specimen to form the taste of youth,

youth, early, to a love of elegance in anatomy, and to ſhow them how much ſhorter, more eaſy, and agreeable it is, to be introduced to that ſcience by means of figures, than by tedious ſyſtems, and lectures upon dead bodies alone. The work was alſo undertaken for the amuſement of the editor at his leiſure hours, who is a profeſſed lover of anatomy, and of the arts of deſign. It was likewiſe undertaken from a veneration of the great Albinus, in order ſtill further to ſpread the knowledge of his admirable works, ſo juſtly deſerving to be known; but particularly to recommend the regular uſe of figures in anatomy, and the true manner of applying that ſcience to the uſes of the arts of deſign. The ſmaller form was choſen, to make the tables more portable, more fit for ſtudy, and at the ſame time to come at a lower price. Tho' ſmall, they are of ſuch a ſize as to contain, in the diſtincteſt manner, whatever is expreſſed in the large originals from which they were copied, ſome entirely, and others partly with my own hand; and engraved, under my own eye and conſtant direction, by a young engraver, who I hope will one day be eminent in his profeſſion. No time or expence was ſpared to give them all the perfection, that copies of ſuch inimitable originals are capable of. The back grounds were omitted, not only to ſave labour and expence, but as tables of ſo ſmall a form did not ſo much require theſe ornaments; and by want of them, the figures ſeemed to appear with more diſtinctneſs and perſpicuity, and to be fitter for the uſe of ſcience. To the outlines or linear figures, on account of the ſize, I was obliged to add figures of particular parts as large as the originals; otherwiſe I could not have found room for the letters or marks of reference; this I hope will be thought a good contrivance, and will not be inconvenient to the reader, the ſeparate parts being placed all around, near the correſponding members of the entire figure, and as it were in the ſame attitude and direction, ſo that the eye paſſes eaſily from the one to the other; and what letters are not found on the entire figure, muſt always be looked for on the ſeparate correſponding parts. And it is hoped that

very

very few errors will be found, even in the linear tables, and letters or marks upon them, which were examined with the ſame attention as every other part of the work; and indeed, in my care of printing the tables, choice of the workmen, and of the paper, and in every other particular, I followed, as nearly as I was able, the excellent method purſued and deſcribed by Albinus himſelf.

The tranſlation, both of the general preface of Albinus, containing the hiſtory of the work, and alſo of his explication of the tables, is entirely new; in which I have not only endeavoured to expreſs the ſenſe, but alſo the graces of Albinus; and in the index, or explication of the tables, his elegant brevity. I had too much pleaſure in endeavouring to imitate that great anatomiſt, and to try to expreſs the beauty and elegance of his manner, to weary of the laborious taſk of tranſlating him anew; eſpecially as the former Engliſh tranſlator, beſides miſſing almoſt every where the character and elegance of the author, is erroneous in many places, and in ſome pages of the hiſtory of the work hardly to be underſtood; chiefly becauſe the tranſlator ſeems not to feel the beauty of the original, and to be totally ignorant of the painter's art, ſome knowledge of which is ſo neceſſary to one that undertakes a work of this kind. I have taken the liberty to divide Albinus's hiſtory of the work into chapters and ſections, and alſo have added an epitome of it; for tho' nothing can be more methodical, and more worthy the ſtudy both of painters and anatomiſts, than that hiſtory, in order to judge of the merit and defects of anatomical figures, and in what manner they ought to be conſtructed; yet as the nature of the ſubject, and the minuteneſs of the author, require an attentive reader, I thought theſe ſmaller helps, by rendering every thing more clear and eaſy, might be uſeful to young painters and anatomiſts. I confeſs however, notwithſtanding all the pains I have beſtowed, that my copies, both of the tables and of the words of Albinus, are many degrees inferior to the originals; but I flatter myſelf they are leſs unworthy of them, than ſome former attempts; tho' perhaps others may diſcover errors and defects in my tranſlation and copies, that I myſelf am inſenſible of. Such errors I ſhall ever be ready to own and to correct; and I ſhall proceed to give the remaining muſcular tables of Albinus, ſo as to complete the work; likewiſe other anatomical tables and treatiſes, according as I find they will be agreeable to the public.

As an ornament to this little work, I have added the anatomy of Celſus, to ſerve for a ſpecimen of that fine author; with intention to ſhow, in how ſhort a compaſs ſo important and difficult a ſubject as anatomy may be treated, in a clear and elegant manner; if my tranſlation of this part of Celſus is approved, I ſhall afterwards exhibit other parts of that accompliſhed author in an Engliſh dreſs. I have likewiſe added a tranſlation of the anatomy and phyſiology of Cicero; in order, if poſſible, by ſo bright an example, to recommend the ſtudy of uſeful and elegant learning, to the great of our own age and country, and thereby revive the manners of greater and more virtuous times. Laſtly, I have ventured to premiſe a ſhort introduction to anatomy, in a manner ſuited to the uſe of the lovers and practiſers of the arts of deſign; and as the attempt is ſomewhat new, I hope its faults and imperfections, with the others found in this work, will meet with the favour and indulgence of learned and candid judges.

Great Titchfield-Street.
January 1. MDCCLXIX.

INTRODUCTION TO THE TABLES,

GIVING

A SHORT VIEW OF PICTURESQUE ANATOMY.

ANATOMY, like many other parts of learning, has often been deſcribed with too much minuteneſs, ſo as to make it tedious and diſagreeable, even to the lovers of that art. Great judgment and ſkill are required to reject the uſeleſs, and to retain, arrange, and adorn the uſeful parts of ſcience, and apply them to practice, ſo as to be agreeable to men of genius, and fit for the generality of mankind, leaving the minute and leſs uſeful things to the ſtudy of the curious. In this method of treating every ſubject, the works of the ancients afford the moſt admirable models, while the bulk of modern ſyſtems, tho' rich in matter, too often confound us with a load of ill digeſted particulars, heaped together without taſte or judgment, and deſcribed without perſpicuity or elegance. A great reformation in this is therefore much required, and on no ſubject more than anatomy; ſo that notwithſtanding the many large ſyſtems and abridgments already publiſhed, a ſhort and elegant ſyſtem of anatomy is ſtill greatly wanted, fitted for general uſe, and for men of liberal education, and particularly for the practiſers and lovers of the arts of deſign: and indeed it would appear, that a method chiefly by good figures and explications, would beſt of all anſwer that purpoſe: for what more natural, ſhort, and agreeable way can be deviſed, to explain the mechanical ſtructure of any machine, than by preſenting it to the eye in a ſeries of proper figures? nor could the fabrick of the human body be ſooner or more agreeably learned, or deeper fixed in the memory, than by copying the beſt anatomical figures, and indeed many other parts of knowledge might be ſooner and more agreeably taught by the aſſiſtance of drawing, than by any other method; for which reaſon, drawing ought to be an univerſal piece of education, and conſtantly taught along with writing, which is only a ſpecies of it. On theſe accounts I have often wiſhed, not only that anatomy were commonly taught in a very different manner from that now in uſe, and that figures were better underſtood, and more uſed, but alſo that ſome learned, judicious, and elegant anatomiſt, would take the trouble to compoſe ſuch a ſhort ſyſtem as I have deſcribed, attended with a complete ſet of figures properly explained: by which means anatomy would be rendered eaſy, and much more agreeable, and therefore become a more general ſtudy, as a part of liberal education; and not, as it is at preſent, be in a manner totally confined to phyſicians and ſurgeons, and even to them, too often taught in a dull, tedious and diſagreeable manner. I have alſo wiſhed, that ſome perſon ſkilled both in anatomy and in the art of painting, would treat of anatomy in a manner particularly ſuited to the arts of deſign; a thing much wanted by the profeſſors and lovers of theſe arts, and little underſtood by the generality of anatomiſts: and I confeſs, I had once ſome thoughts of trying what I myſelf could perform in that way, but finding that it required more abilities than I was maſter of, to give a complete and regular treatiſe upon that ſubject, and alſo more time than I could ſpare from other avocations, I altered my deſign, and inſtead thereof, by way of introduction to the following tables, before the particular explication is conſulted, I thought it might be of ſome uſe to lay a few obſervations on that ſubject before the reader.

According

According to the views that theſe have who apply to the ſtudy of anatomy, their attention mu be applied to different things, and in a different manner. Thus, according to the preſent ſyſtem c medical education, a phyſician muſt ſtudy anatomy on an extenſive plan, and with very enlarge views, ſo as to underſtand not only the larger parts, and groſs mechaniſm of the animal, but alſ to penetrate into its moſt intimate ſtructure, ſo as to diſcover, if poſſible, the moſt minute veſſels cells, pores and fibres, upon which the various ſunctions of the animal depend, and which are th ſeats of particular diſeaſes, or by means of which, remedies may be applied to the whole body, o its particular parts: nor muſt he underſtand the ſolids only, but alſo the fluid parts, which nouriſ the former, and are themſelves the ſeats of diſeaſes, and act upon the ſolids ſometimes as poiſons and ſometimes as remedies; nor ought the finer parts by which the body is governed, and even the mind itſelf, ſo far as it acts upon and is connected with the body, to be leſs the ſubject of medical ſtudy than the body itſelf; otherwiſe, a phyſician muſt have very imperfect ideas of his profeſſion, and of the animal machine, and often fail in his cures, becauſe many diſeaſes are wholly, or partly cured by the movements of the mind, or by applying the remedy firſt to the mind, and thereby producing the wiſhed for effect upon the body. And in like manner, the whole extent of nature, in ſo far as it can any way influence or affect the animal machine, either to injure or reſtore it, is alſo the true and neceſſary ſubject of medical ſtudy; from all which may be ſeen, the importance of the medical profeſſion, and the great extent and difficulty of it, eſpecially as ſo much judgment, honour, humanity, and induſtry, are conſtantly required in the practice of it; otherwiſe, opportunities muſt be loſt, and the greateſt miſchief done; and an art deſtined for the ſafety and protection of mankind, be converted into the greateſt curſe to ſociety. But to return to anatomy.

A ſurgeon on the other hand, tho' he ought to have at leaſt a general idea of the animal œconomy, and indeed of every part of medicine, yet his chief anatomical ſtudy ſhould be confined to know exactly the bones, with their joints, and the muſcles, together with the large blood veſſels and nerves, and the ſituations and mechanical ſtructure of theſe parts, which are to be the ſubject of, or ought to be ſhunned in performing operations, or are the ſeats of chirurgical diſeaſes, or to which external remedies are moſt properly applied.

But a painter, or a lover of the arts of deſign, muſt ſtudy anatomy with other views. As the repreſentation of the outſide or ſurface of the human body, is the chief object of his art, he ought to ſtudy the ſtructure of the body and its inward parts, chiefly for the ſake of, or as they affect or are referred to the external ſurface, and make their appearance there, or are aſſiſtant in the better drawing and repreſentation of it. Hence the parts which ſhow themſelves upon, or affect the ſurface of the body, ought to be the ſole or chief object of the ſtudy of a painter. The parts therefore that lie neareſt to the ſurface or outſide of the body, and conſequently that are moſt immediately concerned in forming its outline, are firſt to be conſidered by a painter, viz. the external layer of muſcles, eſpecially the larger ones, and theſe that] are moſt ſubject to appear in the movements and attitudes of the body: as to the ſkin and fat under it, theſe are uniformly ſpread over the whole body, and are to be conſidered merely as a drapery or covering to the more inward parts, which appear every where more or leſs thro' them, at ſome times and places in a ſtronger, and at others in an obſcurer manner. But tho' the parts neareſt to the ſurface, are the firſt and moſt obvious that belong to the ſtudy of a painter, yet nature has ſo contrived the human body, that the external parts cannot be well underſtood, without a juſt idea of the internal ones, even of thoſe which are as it were buried in the center of the body:

I mean the bones, or ſkeleton, which are the foundation and frame on which the whole fabrick is built, and to which, as a baſis, all the other parts are mediately or immediately referred, particularly the muſcles, ſo neceſſary to be known by painters, which are chiefly inſerted into the bones, and make conſiderable marks and impreſſions upon them; and conſequently, without the knowledge of the bones, the muſcles and other ſoft parts cannot be underſtood: but there is another reaſon why the bones muſt be ſtudied by a painter, viz. becauſe parts of the bones, tho' covered by the integuments, appear not obſcurely to the eye in many places of the body, and like the large muſcles, are there the cauſe of the outline, and of the character, proportion, beauty, and appearance of many parts; and when properly conſidered and underſtood, the bones, by ſo many fixed points, give the fineſt direction to a painter, not only how to find and place the muſcles, but alſo how to draw the human body; nor can it be ſo juſtly or readily drawn by any painter, as by one that underſtands anatomy in a maſterly manner, and particularly the bones and external muſcles, and can point them all out upon a living man, and by means of that knowledge, determines all his points, and the forms and proportions of every part and member, adding one part to another as he knows they lie upon the body: this is the true and natural method of drawing the human figure, and is a much eaſier and compleater way, to one that underſtands anatomy, than any artificial or mechanical method by ſquares, or by dividing the body into ſo many heads, or by truſting merely to practice and memory, or a ſervile imitation of any maſter. But tho' the bones and external muſcles are the moſt neceſſary part of the anatomical ſtudy of a painter, yet it muſt be confeſſed, that at leaſt a general knowledge of the whole fabrick is of great uſe, in order to a more complete and maſterly repreſentation of the human body, and in order to be able to diverſify, and give a reaſon for every appearance; and not only the ſolids muſt be known by a painter, but he ought likewiſe to have ſome idea of the fluids, as on theſe chiefly depend the various tints and colours of the ſkin, that appear in the different ſexes and ages of life, in different characters and occaſions, climates and nations, even to that of the Blacks or Æthiopians. And as nature has ſo contrived the human frame, that the movements and paſſions of the mind affect the body, and are evidently ſeen and diſtinguiſhed upon the countenance, and are expreſſed there and in other parts of the body by ſtrong and certain characters, and as this is the moſt delicate and higheſt part of the painter's art, by which he is capable to move, to delight, and to inſtruct mankind, and to recommend himſelf and his art to their eſteem and admiration; therefore, the ſtudy of the mind, and its various characters, paſſions, and movements, in ſo far as they are marked upon, and expreſſed by the body, ought to be above all things the ſtudy of a painter: for as the members of the human body, in a good picture, beautifully appear thro' the drapery, and as the bones and muſcles appear thro' the ſkin, ſo the mind itſelf in all its characters and paſſions appears upon the countenance, and in the expreſſive proportions, attitudes, and tints of various parts; by which, as in a pantomime or dumb repreſentation, a painter can as it were ſpeak to the beholders, and by lines and colours alone, can perform the ſame effects with the muſician, the poet, the orator, or the actor upon the ſtage of mimic, or of real life.

A lover of the arts of deſign, or indeed any anatomiſt of true taſte, will look upon the human body and all its parts with the eye of a painter, otherwiſe, he will ſee and deſcribe it in an ignorant and ruſtic manner: this pictureſque turn we obſerve in few modern anatomiſts, but rather a great ignorance of it, the generality ſeldom riſing above mechanical ideas, and many of them have even been ignorant of geometry, and every polite and liberal ſcience, though abſolutely neceſſary to a true knowledge of anatomy. Obſerving the human body with the eye of a painter, enables us to ſee it in all its beauty and perfection, and raiſes in our minds a thouſand ideas of the uſes and propriety of the ſeveral parts, whereof one ignorant of painting will be totally inſenſible: and in deſcribing the human body upon this plan, we naturally do it in the moſt clear, ſhort, and agreeable manner, far different from the dull pedantic deſcriptions and tedious trifling of vulgar anatomiſts. It is from bad habits alone, and mere want of genius, that any noble ſcience, or any deſcription of nature, can become tedious or diſagreeble, or be born and reliſhed by the hearers: hence the works of the ancients, and of thoſe who follow their footſteps, are read and ſeen with delight and admiration, while we are apt to fall aſleep over the works of many accurate and laborious modern writers, and wonder how men can be ſo blind and inſenſible to true beauty, when nature and ſuch admirable models are conſtantly before their eyes.

Having premiſed theſe few obſervations, I might remit the reader to the tables themſelves, with their explication; by the careful peruſal of which, a tolerable idea may be formed of the ſkeleton and external muſcles, at leaſt, for the uſes of painting, but the young anatomiſt and painter will perhaps better underſtand them, and with more eaſe and pleaſure, and be able more fully to connect the particulars, and apply them to the arts of deſign, by means of the following ſhort ſketch of pictoreſque anatomy, which, in its turn, will alſo be better underſtood by conſulting all along the tables and the explication, this introduction and the tables mutually tending to explain and illuſtrate each other.

OF THE SKELETON.

The ſyſtem of the bones or ſkeleton, is as it were the ſolid frame that contains, defends, and gives ſtability to the ſofter parts, and to which they are ultimately attached; and conſequently this bony fabrick has of itſelf the general form, ſize, and appearance of the entire body (Tab. I. II. III.). This ſolid frame is moſt artfully compoſed of different parts jointed one to another, ſo as to be capable of every uſeful and graceful motion, in the whole and in all its parts; and the various bones and pieces of which it is compoſed, differ in ſize, form and ſtrength, in poſition, connexions and motions, according to the uſes and exigencies, and even the beauty of every part, to which they often add a certain grace and character, by obſcurely appearing here and there through the ſoft parts, even in the living body. The head, which the painters conſider as an oval. (Tab. I. II. III.) is, as it were, the dome or cupola to the whole edifice. In this higheſt part the ſenſes are placed, and the brain defended by ſolid bone; the head, like the reſt of the body, derives its ſize, form, proportions, and principal characters firſt from the bones, but the ſoft parts that cover them add the life, the motions, and the finiſhing beauty, in which laſt, the hair alſo concurs; and it is ſurpriſing how ſo few ſimple organs, and ſo thin a covering of ſoft parts, are capable of ſuch infinite variety of forms and expreſſions as we ſee in the human countenance, affording an endleſs field of ſtudy. In the head the bony part is a more complete fabrick, and comes nearer to the form of the entire body than in any other part of the ſkeleton; and being the ſeat of ſo many noble organs, and the chief part to be ſtudied by a painter, it deſerves the firſt place, according to the common cuſtom of anatomiſts. Here veſtiges of the ſmooth poliſhed bone ſhew themſelves on the forehead, in the riſings all around the eye, in the hollow of the temples, on the noſe and cheek bones, and margin of the lower jaw, giving great pleaſure to a painter that underſtands anatomy. Next comes the elegantly bent pillar of the ſpine, (Tab. III.) ſtrong, yet flexible, by conſiſting of ſo many parts firmly tied together. This bony column, at the ſame time, gives ſize, ſtrength, and motion to the body, attachment to many ſurrounding parts, and being hollow through its whole length, ferves to conduct and fecure the fpinal marrow, and to tranfmit nerves to every part of the trunk and extremities. The fpine confifts of four and twenty vertebræ, (Tab. I. II. III.) generally increafing in fize as they defcend, and gradually varying in their figure: feven of thefe vertebræ belong to the neck, which admit of peculiar and confiderable motions, and allow of many graceful movements to the head and neck. The next twelve belong to the back, thefe are almoft rigid, and admit of very little motion; to thefe, as to a folid bafis, the twelve ribs of each fide are attached, which together with the fternum, and their own cartilages, form a kind of yielding cage or bafket, to contain the heart and lungs, (Tab. I. II. III.). This bony cage admits of a fmall motion when we breath; and to the lower margin of it all around, is fixed the diaphragm, a tranfverfe mufcular partition, dividing the thorax from the abdomen, a main organ of refpiration and of other functions. The five lower vertebræ belong to the loins, and admit of confiderable motion, of great ufe in the firm and graceful attitudes and flexions of the trunk, and in many offices of common life. Between the ribs and pelvis there is a great void in the fkeleton, efpecially before, (Tab. III.). In this fpace lie many of the abdominal vifcera, with the parts that contain and cover them, making on the forepart the beautiful fwell of the abdomen, elegantly marked by the containing parts (Tab. IV.). To the fuperior part of the thorax, by means of the tranfverfe clavicles and of large and ftrong mufcles, are appended the upper extremities, which at the fhoulders give breadth to the thorax above, and ferve many noble purpofes of ftrength, of art, of defence, of expreffion, and of beauty. Thefe are divided into the fhoulder, confifting of the clavicle before, and the thin broad fcapula behind, which moving free among the mufcles, by their means governs the motions of the whole arm; and its triangular form has a moft beautiful effect, feen floating among the foft parts in the naked figure (Tab. V.): and indeed the whole fhoulder is a moft noble part, and a fine exercife to a painter that underftands anatomy, for befides many fine large mufcles, the bones themfelves alfo moft beautifully and diftinctly appear. Next comes the arm bone, capable of a large and free motion, whofe round head at the fhoulder in lean perfons obfcurely appears, and at the lower end its condyles are evidently feen, where it is joined to the forearm; this confifts of the radius and ulna, which move upon the arm bone with the more confined motion of flexion and extention, but for the fake of the hand and its various and important ufes, the radius and ulna likewife revolve upon each other lengthways, in a very curious and fingular manner, turning the hand alternately prone and fupine, as upon an axis. Laftly comes the hand itfelf, the moft fimple and curious machine in nature; it confifts of the carpus, metacarpus, and five fingers, the thumb being as it were an antagonift to the other four; the whole by its general form, and different parts and motions, ferving almoft every poffible ufe, and its various attitudes being capable of great beauty and variety, an infinite field to painters, and moft worthy of their ftudy, and indeed, next to the countenance and the voice, the moft beautiful and expreffive part of the human body.

We come now to the pelvis and lower extremities. The pelvis fupports and defends the lower vifcera. The back part or os facrum (with its coccyx) of a triangular form, is at it were the bafis and continuation of the fpine, whofe vertebræ it obfcurely refembles, and performs its offices, by receiving the extremity of the fpinal marrow, and tranfmitting nerves to the furrounding parts. The lateral and foreparts of the pelvis, though fixed and immoveable, anfwer in fome refpect to the fcapulæ and clavicles, as they afford fockets for the thigh bones, and alfo a feat to many ftrong mufcles that belong to the trunk and extremities. The upper margins of the offa ilium appear gracefully in the living body on the forepart, and form a kind of boundary between the belly and the thighs. The fpines of the os facrum, as of the vertebræ, obfcurely appear in bodies not loaded with fat, as alfo the great trochanter of the thigh, the reft of which bone, till you come to the

the knee, is deeply immerfed among large and ftrong mufcles: but at the knee, the bones make a very fine appearance, viz, the condyles of the thigh bone, the tops of the tibia and fibula, and the round patella, (Tab. I. II.) a bone fo beautiful and fo ufeful in the government and defence of this joint. Here fkilful painters and fculptors never fail to fhew their art, not only in the entirely naked figure, but in fome ancient roman habits, in which this beautiful joint appears, and indeed the ancient dreffes, and even fome of the gothic ones, greatly excel the modern, as they not only cloath, but adorn the human body, fhewing its feveral parts to advantage, and giving a noble field to painters and fculptors, who, when they want to add dignity and beauty to their figures, are obliged to borrow from the dreffes of the ancients, as we do from their languages, architecture, and the other arts of antiquity. The bone of the tibia appears through the whole length of the leg, and at the lower part of the tibia and fibula, the two ankles elegantly appear, and fix the bounds between the leg and foot. The foot, a thick and folid part, ferves as a bafis and fupport to the whole body, and therefore its parts are only capable of obfcure motion; it confifts of the tarfus, metatarfus and toes: in the whole, and in every part, it in fome fort refembles the hand; and tho' much inferior, comes next to it in beauty, and therefore great artifts feldom cover this part, but like the hand, take pleafure in fhewing it naked in all its varieties.

The various conformation of certain parts, chiefly at the extremity of bones, is the principal caufe of all the variety of the joints; which are compleated by means of ligaments that bind them together, and of fmooth cartilages, and a certain lubricating moifture to enable the articulated parts to flide quickly, fmoothly, and gently upon each other. By means of the joints, the human body becomes a moving fabrick, a thing neceffary in the common offices and arts of life, alfo for health, defence, and amufement. By the joints, moft of which fo elegantly appear to the eye, the body is not only fubdivided into a multitude of well proportioned parts and members, compofing one harmonious whole, beautiful to the eye, but is thereby capable of an infinite variety of ufeful, expreffive, and graceful attitudes and motions; and though every joint has its peculiar ufe and extent of mobility, determined by the nature and conformation of the parts that compofe it; yet the joints, as we fhall fee afterwards of the mufcles, feldom act feparately and alone, but like thefe, beautifully co operate one with another, in all the principal attitudes and movements of the body; fo that in many pofitions, almoft all the joints, as well as principal mufcles, are more or lefs concerned, and act in harmony one with another, having each a certain fhare in thefe ufeful and beautiful movements. But a particular defcription of the different joints, with fuch obfervations on them as properly belong to the painter's art, though a moft curious and ufeful part of anatomy, would be too tedious for a fhort introduction of this kind.

Though there are only male and adult fkeletons reprefented in thefe figures, we may obferve, that the difference of fexes and of ages is feen even in the fkeleton, as well as in the entire body; not to mention the difference of ftature, and of the fize, ftrength and form of particular bones, even to the fingers, the different proportion of the fhoulders and pelvis in the two fexes is remarkable. In the male, the fhoulders are broader and the pelvis more narrow; the contrary is the cafe in the female fkeleton: and befides, the whole has a more feminine appearance, the bones are fmoother and more delicate, with much lefs roughnefs from the impreffions of the mufcles and furrounding parts. The like may be obferved of the fkeletons of children, the whole of which have the fame appearance, and the parts the fame proportions with the correfponding parts of the entire child. The large globofe head, the round face, the fhortened trunk and extremities, the bones thick, foft,

and

and almoſt every where imperfect, the proceſſes, protuberances; and marks or impreſſions, leſs evident, and the bones conſiſting of many parts and diviſions, which are afterwards united in the hard and perfect bones of adults.

But the differences of age and ſex are not the only ones we can perceive upon the ſkeleton; a ſkilful perſon can eaſily diſtinguiſh the ſkeleton of a well made delicate body from that of one of a more rude or homely make, and all the different degrees of deformity: we may with Veſalius diſtinguiſh the ſkulls of different nations; and in like manner, by a more nice and accurate attention, we may go a great way to obſerve almoſt every character and diſtinction, that is perceivable in the entire man; for the impreſſions of early habits, which from the mind, or from other quarters, affect the general appearance of the entire body, generally communicate themſelves even to the bones, which being long in a ſoft and growing ſtate, are evidently ſuſceptible of changes and impreſſions of many kinds, from the nature and action of the ſurrounding parts.

So much for the bones and their joints, in ſo far as may ſuffice to give a faint idea of theſe parts for the uſes of painting and ſculpture, and to ſerve the lovers of theſe arts as an introduction to the explication of the following tables. We now proceed to give a like idea of the muſcles.

OF THE MUSCLES.

The ſkeleton is one ſimple ſyſtem of ſolid parts, ſeen as it were at one view, and ſerving as a jointed frame on which to build the reſt of the body. But the muſcular or fleſhy parts, that cloath and move the ſkeleton, are ſoft, and form a more various and complicated ſyſtem, conſiſting of different ſtata or layers, one covering another, and divided into numerous portions of different ſize and figure, regularly diſpoſed over the whole body, compoſing a great part of its bulk, and the chief cauſe of the ſize and form of the members; for when ſtript of its uniform coverings, viz. the ſkin and cellular or fatty membrane, the external muſcular figure nearly reſembles that of the entire body.

The muſcles differ greatly in their ſize, figure, and other particulars, according to the parts where they are ſituated, and the uſes to which they are applied. But in general they are compoſed of fibers; the middle part or belly being large, ſoft and red, and the extremities or tendons, which are generally inſerted in bones, being ſmaller and harder, white and ſhining; the red part is properly the moving power, and acts by contraction, during which it ſwells, becomes hard and ſhorter, ſometimes to a great degree, and thereby pulls the parts to which its extremities are affixed. The muſcles are governed by the power of the will, except the fibers of the heart and of the inteſtines, which of all others are moſt irritable; the muſcles of reſpiration act in both ways. The muſcles can act in the moſt gentle and delicate manner, and alſo with great ſtrength and velocity, though much of their power is loſt, by the places and manner in which they are often ſituated and inſerted on the parts to be moved. The cauſes of muſcular motion, are difficult to be accounted for from the known ſtructure of muſcles; great velocity communicated to the nervous fluid by the mind, ſo as to ſtimulate the fibrils, ſeems the moſt probable account. The muſcles are arranged in their places, and allowed to ſlide upon each other, by means of the cellular or fatty membrane, and their fibres are lubricated every where by the oil it contains; and in the fabrick of the body, and of the muſcles themſelves, many contrivances are uſed to aſſiſt their actions. The muſcles are in ſufficient number, and ſo diſpoſed and contrived, as to be a warm covering and defence to the more inward parts, as well as to move the joints in all the directions they are capable of, to aſſiſt in many functions of the body, and to place and retain it in every poſſible attitude; in doing which, the particular muſcles ſeldom act alone, but in the moſt various manner co-operate with or oppoſe each other: ſo that the whole muſcular ſyſtem may be conſidered as one muſcle, every fibre being entirely under the power of the will, at the pleaſure of which, the whole body and all its parts are at once or alternately moved and governed, as it were by ſo many bridles. Beſides this grand purpoſe of the muſcles, they likewiſe ſerve the general uſes of the animal machine, being the chief cauſe of reſpiration, and of the circulation of the blood and juices, alſo promoting digeſtion, abſorption, ſecretion, excretion, nutrition and growth; hence they likewiſe prevent and cure obſtructions and other diſeaſes, and by their inceſſant action are one great cauſe of the hardening and waſting of the body, and the decays of old age.

This is a general idea of the muſcular ſyſtem; but a painter muſt ſtudy it with particular views to his own art. He muſt conſider, that the muſcles chiefly form the ſize and outline of the body; that many of the external muſcles have regular forms, and beautifully appear at all times under the ſkin, but eſpecially when in violent action; that in that caſe, even deep ſeated muſcles ſometimes appear, as alſo more clearly the bones and other parts; that the different parts of the ſkeleton bearing obliquely upon each other, and upon the feet or common baſe, even in the moſt ſimple attitudes, many muſcles are therefore conſtantly in action, ſucceſſively relieving each other, in order to preſerve the æquilibrium, changing their actions and appearances on the ſurface of the body, as the various poſtures, attitudes, or exertions require. Theſe appearances he ought diligently to obſerve, even in different bodies, and compare them with his knowledge of anatomy, in order to apply them juſtly on proper occaſions to adorn his figures. In this the ancient artiſts far excelled the moderns; and indeed, not to mention their other advantages, they had better opportunities of obſerving the naked body in the gymnaſia, when employed in the manly exerciſes of the palæſtra. But why are not theſe inſtitutions revived along with the other noble diſcipline of antiquity, founded on nature and reaſon, and aiming at the perfection of the human kind? If this were done upon a proper plan, as our lights and opportunities are ſuperior to theirs, we might in time not only equal, but even excel them. Is it not the diſgrace of a nation, that ſo much ſtudies and admires, and above all others reſembles the antients, and ſeems fond to rival them, to be ſo long ſunk in narrow, partial, and ſelfiſh ideas, that retard its glory and progreſs to true greatneſs? Not only men of genius, but the people in general long for, and are prepared for this deſirable event, and wait only for leaders worthy of them. The ancients were perfect maſters in applying anatomy to the arts of deſign, they not only knew the general form and places of the muſcles, but how to vary their appearance in every degree of action and of character. The muſcles of a Hercules, for example, differ from theſe of an Apollo, and of an Apollo from theſe of a Venus, in the ſame character and ſtile as the figures themſelves: the muſcles of the dying gladiator ſeem to die along with him, and in the fighting one, and the wreſtlers, they are agitated like the figures themſelves, and the parts to which they belong. In the Laocoon they ſeem to be convulſed and trembling. In beautiful bodies they are beautiful, as they ought to be, but in the deformed, as in Silenus, the muſcles are deformed like his whole figure, and ſo in other varieties; whereas in many modern works, not only theſe judicious and delicate expreſſions are unknown, but the greateſt ignorance in anatomy often appears, either by falſe repreſentations, or by a dull and injudicious oſtentation of anatomical knowledge, on every occaſion; the ſame muſcles appearing, and almoſt in the ſame character in every figure, and either inanimated like the ſimple diſſection of a dead body, or ſwelled and contorted in an extravagant manner; while ſome more prudent

prudent, and conſcious of their ignorance of anatomy, repreſent the human body like a ſkin ſtuffed with wool, without any marked diſtinction of bone or muſcle: others are totally ignorant of the co-operation of muſcles, and how to allow their general effects to appear, without bringing the particular muſcles to view, as in many fine expreſſions about the eyes and mouth, and in other parts of the body. So that after being perfectly maſter of common anatomy, much ſkill and judgment is ſtill required, in order to apply it properly to the arts of painting and ſculpture, if an artiſt has ambition to pleaſe judges above the vulgar.

Many more obſervations might be made, with regard to the general ſyſtem and anatomy of the bones and muſcles, and other parts of the human body, with the application thereof to the arts of deſign, did the bounds preſcribed to this ſhort introductory ſketch allow of it. Reſerving theſe therefore for ſome other occaſion, we ſhall conclude what we have to ſay of the muſcles, with a ſhort view of the external layers, as they appear in the three muſcular tables now publiſhed; confining ourſelves chiefly to the uſe of painters and ſculptors.

In order to underſtand theſe figures aright, the three muſcular tables muſt be conſidered and compared together as one, under the idea of a ſolid figure, which can be turned round and preſented to the eye in different views; eſpecially the front and back, which are in the ſame attitude, and contain in themſelves the entire round of the body. And in the like manner may be compared the three views of the ſkeleton. This being done with care and attention, each ſkeleton muſt be compared with its correſponding muſcular figure; and both the one and the other, with the entire living body placed in the ſame attitude; by which a tolerable knowledge will be acquired of the anatomy of the human body, in ſo far as it belongs to the arts of deſign.

Though much might be ſaid for the uſe of painters, not only on the different ſtrata or layers of muſcles, but alſo upon particular muſcles; yet at preſent we ſhall confine ourſelves chiefly to the external layers that appear in theſe tables, and which lie immediately under the ſkin and cellular membrane, and retain ſo much the entire figure or outline of the body, that painters generally inſcribe theſe muſcles upon that outline as a baſis. However, as ſome deep ſeated muſcles on ſome occaſions more or leſs appear, at leaſt by their effects, particularly, the diaphragm that as an antagoniſt ſuſtains the beautiful ſwell of the abdomen, and the muſcles that ſupport the trunk and govern the ſpine, alſo the muſcles concealed in the orbits of the eyes, ſo uſeful in expreſſing the paſſions and movements of the ſoul; I ſhall only juſt mention that ſuch muſcles exiſt, and ought to be known by painters, though they do not appear in theſe tables. But as to the layers or ſtrata of muſcles, and particularly the external one, we may obſerve in general, that though they might have effectually ſerved the purpoſes of moving powers, and indeed all the other uſes of muſcles, by being formed of leſs regular figures, and placed on the body with leſs exact order and compoſition; yet nature, conſulting grace and beauty as well as utility in all her works, has ſo contrived the muſcular ſyſtem, that, while it effectually performs its ſeveral functions, not only the particular muſcles are formed with great variety, into beautiful and regular figures, of a ſize and appearance ſuited and proportioned to the ſeveral parts, but the whole together is ſo diſpoſed, as to exhibit an agreeable compoſition to a lover of anatomy; and they are ſo placed and ſecured, that in their moſt violent actions, they cannot ſtart up ſo far from their true places, as to hurt or deface the form of the body or of its ſeveral parts; but rather by their gentle ſwellings and depreſſions, tend to encreaſe its beauty; and to gain this end, nature has made a great waſte of muſcular ſtrength. Of the different ſtrata of muſ-

cles,

cles, the external one reprefented in thefe figures is more beautiful than the hidden and internal ones, becaufe it confifts of larger and more regular maffes than the internal ftrata, and as the whole comes nearer to the form of the entire body. Again, of the three external views that of the front is moft beautiful, not only on account of the face, and becaufe the limits of the trunk are more exactly determined both above and below, and as the extremities feem more to belong to the foreparts, and are there more beautiful; but alfo, becaufe in the front there is more variety and elegance in the mufcular appearances. The back view, though in fome refpect more rude and fimple, likewife has its beauty, efpecially about the neck and fhoulders; alfo about the hips and the whole lower extremities, whofe bones and mufcles are larger and ftronger than thefe of the fuperior, fuited to the fize and office of the parts. The profile or fide view, both in the fkeleton and mufcular figure fhews many particulars that cannot be fo well perceived in the others; for example, the direction and bearing of the parts one upon another, and how far they project both before and behind; as the head over the neck, the various curvatures of the fpine, and fo of other parts; it alfo fhews fome parts more fully, and others in a more pleafing attitude; and from the profile, you may likewife compare the proportions of the narrow lateral view of the body, to thefe of the front and back.—In this lateral view, fhould not the right arm have been more elevated backwards, (fee Tab. III. VI.) in order to fhew the internal parts of that arm, as fully in the mufcular figure as they are feen in the fkeleton?

By thefe figures we perceive, that the number and ftrength of the mufcles, are every where fuited to the feveral parts they ferve, and the joints they move. Thus the mufcles on the trunk are few, but generally large, broad and flat, ferving for walls and coverings as well as moving powers; but on the extremities the mufcles are numerous, and moftly oblong, fuited to the fize, form, and action of thefe parts, and the many joints to be moved. About the fhoulders and hips, the mufcles are fhort, large, thick and ftrong, giving the idea of that ftrength neceffary for ftrong and violent action, and to command large members that are fo conftantly in motion: whereas about the fingers, which befides the ftrong, are alfo intended for arts and delicate movements, we find many fmaller mufcles; and ftill more fo on the face, and about the organs of the fenfes and of the voice. Laftly, the deep feated mufcles that command the fpine, run as it were parallel to that bony pillar, and are beautifully fubdivided like the fpine itfelf, fo as equally to govern every part of it.

As to the defcription and ufes of the particular mufcles, efpecially of thefe that afford the moft ftrong and beautiful appearances, not only in thefe three views, but in every different attitude and action of the body, we cannot enter into it in this fhort introduction; but the principal things may be underftood by what has been already faid, and by examining thefe, and other good figures, and comparing them with nature and the works of the beft artifts: and in cafe this little effay is approved, fuch particulars, attended with proper figures, may be added on a future occafion. In the mean time we may obferve, that of the mufcles that are feen evidently in thefe figures, the moft remarkable are; on the head: the mufcles of the face that govern the features, the temporal and maffeter that move the lower jaw; on the neck: the fternomaftoïd of each fide before, covered by the latiffimus colli, that fo beautifully fhow themfelves in the motions of the head and neck; on the fhoulders: below the clavicles, the pectoral and deltoïd; and behind, the large triangular cucullares, that chiefly fupport and govern the fcapulæ, reaching to the head; on the trunk: the abdominal mufcles, and the beautiful indentations of the great ferrati; and behind, the latiffimus dorfi, and feveral mufcles that lie upon the fcapula, viz. the teres major, teres minor, and infrafpinatus; on

the

the thighs before, appear the recti and vasti that extend the knee, and chiefly support the thigh on that joint when we stand, the beautiful transverse band of the sartorius, and on the upper part the tensor vaginæ, and more internally, part of the internal iliacs, psoæ, pectinei, great and long adductors, that chiefly govern the thigh and support the trunk upon it; on the hips and thighs behind: lie the great glutæi, and below these, besides part of several muscles just now mentioned, appear the muscles that bend the knee, and likewise help to support the thigh and trunk when we stand, viz. on the outside the bicipites, and internally the semitendinosi, on each side of which are seen the semimembranosi, also the graciles; the insertions of several of these muscles are more distinctly seen near the knee in the profile figure, and by comparing all the figures, we may form an idea of the beautiful articulation of the knee and the other joints, the hollow of the ham and axilla, the various depressions between the muscles, the muscles that bend and extend the fore arm, and compose the calf of the leg, many muscles that move the hands and feet, the fingers and toes, with their tendons and ligamentary bands; lastly, the parts of the bones that make their appearance here and there between all these parts. But for the particular description, we shall at present refer to Albinus's own explication of these tables, and other anatomical works.

The three simple views of the skeleton and muscular figure presented in these tables, though in themselves not ungraceful, are chosen chiefly for the uses of elementary anatomy, to shew all the parts successively in a plain and distinct manner, and the muscles are represented as they appear in the dead body, without the imitation of life and action, yet from these figures, well understood and compared with nature, joined to observations and experiments on the naked living body, and on the works of great artists, a skilful painter or sculptor will be able to represent the anatomical appearances of the human body in every other position, and also to add these swellings and sinkings, and other marks that always accompany life and action, especially in lean and athletic bodies; in doing which, though the truth of nature ought to be the general rule, yet certain licences may often be used here as in the other parts of painting, (provided they are conducted with judgment and skill) of representing these appearances rather stronger on certain occasions than they actually are in nature; whereby an artist may not only exercise his genius and invention, but give great pleasure and delight to the truly learned in anatomy and the arts of design. However, ordinary artists should be very cautious not to abuse these liberties, as is but too often the case, and should always take care to have reason, and if possible the authority and example of some great master on their side.

Though the bones and muscles are the chief object of the study of a painter, yet other parts must not be neglected, particularly the skin and the cellular or fatty membrane, and the large veins that appear on the surface of the body. The skin is not only the seat of these tints and colours that on many occasions characterise and adorn the outside of the body, and especially the countenance, but is also the seat of the folds and wrinkles of different ages, and that characterise different parts, and of these that express the passions and movements of the soul (for in the skin many small muscles of the face are fixed), and according as the skin is looser or tighter on the parts, or more or less bound down or supported by the cellular or fatty membrane, the appearances of the parts below them alter every where.

I may likewise add, that painters, but especially these whose particular profession it is to paint the brute animals, ought to be acquainted with at least the general principles of what is called comparative anatomy, otherwise they never can completely express the characters, the beauties and varieties of theſe animals, which is only to be done by comparing their fabrick with the outward appearances preſented to the eyes; and indeed it is a general rule that no ſubject whatever can be truly painted, without underſtanding as a philoſopher the nature and properties of it: for which reaſon, hiſtorical, and even portrait painters, ſhould at leaſt be acquainted with the anatomy of theſe animals which are moſt commonly introduced into their works, particularly of that noble and uſeful animal, the horſe, and of that faithful companion of mankind, the dog: as for other animals, as they more rarely appear in pictures, and are leſs particularly known and attended to, a ſlighter repreſentation may generally ſuffice, eſpecially as nature having cloathed the brute animals with various coverings that hide the inward parts, the anatomical appearances in them are not ſo viſible, nor indeed ſo beautiful as in the naked body of man.

And in general, as to the human body, tho' its outward beauty in all its ſeveral parts and members (which is of itſelf a large and ample field) is one great and neceſſary part of the ſtudy of a painter, yet as theſe outward appearances can neither be perfectly underſtood, nor expreſſed, without a conſiderable knowledge of all the internal parts of the body, and even of the ſoul; therefore a general knowledge of anatomy, and of human nature, along with many other uſeful ſtudies, both in the fields of nature and of art (which I ſhall not particularly enumerate in this place) truly belong to the arts of deſign, and ſhould be known by every artiſt, who is ambitious to underſtand the true principles of his art, and to practiſe it according to theſe principles: nor is this great field of ſtudy ſo tedious or difficult, as at firſt ſight it appears, if purſued upon a proper plan, eſpecially as a great part of it may be learned by drawing and modelling alone, and as the pleaſure and advantages it affords are a conſtant ſpur to the induſtry of the artiſt, who ſhould conſider, that by ſuch methods alone, he can truly excel in his profeſſion, and by ſuch alone, the great maſters of ancient and modern times, were able to arrive at the perfection ſo much admired by true judges.

As to the method of ſtudy: The beſt way to begin that of anatomy, or of any art, is by the help of an able maſter, who perfectly underſtands the ſubject, and the art of teaching it in a ſhort and agreeable manner; after which, the ſtudy of authors and figures will be eaſy and delightful. In anatomy, the beſt authors for a young painter, are Heiſter's Compend, Haller's Outlines of Phyſiology, the Tables of Veſalius and Euſtachius publiſhed and explained by Albinus; but above all, Albinus's original tables of the human ſkeleton and muſcles; to which may be added, the Anatomy of Celſus, ſome parts of Winſlow's anatomical work, and of Albinus's Hiſtory of the Muſcles, alſo his Oſteology, which contains an exact deſcription of the bones, written in an elegant and pictureſque manner. Many other works might be added, by thoſe who have time and curioſity to apply to the ſtudy of this ſcience, but for the generality, theſe I have mentioned may ſuffice.

So much for elementary anatomy, but in order to apply it to the arts of painting and ſculpture, the works of the beſt artiſts muſt be conſulted and ſtudied, both of thoſe who have actually applied it in practice, and of thoſe who have written on this part of the principles of the arts of deſign. Tho' the works of the ancients, as has been ſaid, excel all others in moſt particulars, ſo alſo in the judicious and delicate application of anatomy to theſe arts; yet modern times have produced many learned and accompliſhed artiſts, who have ſhown great genius and ſkill in this as in other parts of their profeſſion. At the reſtoration of painting, Da Vinci was fully ſenſible of the uſe and

importance

importance of anatomy; the great Michael Angelo used anatomy even to excess, but in a bold and manly character, and in this respect may be looked upon as the Vesalius of painters; Raphael, his great rival, like Eustachius, softened anatomy more to the truth of nature, and to the beauty of the antique, giving it at the same time the graces peculiar to his own genius; Hannibal Carrachi is just and masterly in his anatomical expressions, and knew thereby how to give both strength and beauty to his figures. Many other great artists might be named of different characters, in respect to anatomy, as in other parts of their art: thus Rubens was fully master of anatomy, as of every art that could form an accomplished painter, and gave it the richness and strength peculiar to his manner, producing a new and riper æra of the painter's art, which the tables and works of Albinus may be said to have done in anatomy. It were much to be desired, that two such noble arts as painting and anatomy, were always in the hands of such artists as I have mentioned, and like other liberal arts were not too often disgraced by the men who professed them. However, notwithstanding all that has been done by the great in these arts; it were still to be wished, for the sake of learners, as I observed at the beginning of this introduction, that a complete compend of anatomy with figures, fitted to the use of the lovers of the arts of design, was composed by some able anatomist, who at the same time understood the principles of the art of painting, and the works of the great artists both ancient and modern, at least in so far as concerns anatomy, and its use and application to the arts of design.

Before I conclude, I cannot but congratulate our country, on the great efforts that have been made of late years in this capital, to promote and encourage the arts of design in all their branches, which had been hitherto so much neglected in this nation—Arts so useful and ornamental to every people, but especially to a commercial one—Arts which have been the delight of the greatest princes in all ages, and which have flourished along with politeness, or sunk in times of barbarity —Arts which nature so strongly recommends in all her works, by exhibiting to our eyes an endless field of study and delight—Arts by which the great nations of antiquity polished themselves, adorned their cities, and handed down their fame to distant ages, by buildings, by statues, coins, and other monuments—Arts by which modern Italy has attracted the attention and veneration of foreigners— Arts that like eloquence and poetry may be universally applied to every purpose, both of public and of private life, to display and record the wonders and beauties of nature and art, to instruct and to polish mankind, to recommend wisdom and virtue, to punish and ridicule folly and vice, to enoble religion by adorning the temples of the Gods, to add dignity to the state, to record great actions, to honour and reward private virtue, to illustrate sciences, to improve arts and manufactures of every kind, from the greatest to the least, and consequently to increase wealth and commerce. In a word, as visible nature affords to the eyes an infinite field of instruction and delight, in every scene of her works, so art, by following her footsteps, may as it were rival her in new and endless scenes of use and beauty. However, tho' by the encouragement and patronage of the great, by the exertion of the artists themselves, by premiums, by exhibitions, by private schools of design in the capital, and even in distant parts of the kingdom, and by other means, these arts have been of late years greatly advanced, and the taste and attention of the public awakened thereto, yet much is still wanting to establish them upon a complete and regular plan, so as to produce their full effect, suited to the dignity and demands of a great and commercial nation: nor can this great end be accomplished, till education in these arts is conducted on new and more extensive principles, suited to the present state of this age and nation; a grand national academy should therefore be erected in the

the capital, upon the most extensive and generous plan, not only for the arts of design, but for the improvement of every other art and science, and for finishing the education of the noble youth, in every useful and elegant art, that can strengthen and form the body or the mind either for war or peace, for public or for private life; on the model of which, our public schools would soon be obliged to reform themselves: by which, in the next generation, we might expect to see a very different race of men from what we are likely to have, when things are left merely to chance, upon the present system of indolence and dissipation.

P. S. Since this little work was sent to the press, we are informed, that his Majesty has been pleased to erect an academy for the arts of design, in their full extent; a thing long and ardently wished for in this nation; by which it is to be hoped, that the above great and desirable ends, and every other noble effect that these arts are capable of, will in due time be attained by the artists and genius of the British nation, united under the immediate influence of their Sovereign.

THE

SIX TABLES OF ALBINUS,

WITH THEIR LINEAR FIGURES;

ALSO,

ALBINUS'S HISTORY OF THAT WORK,

AND HIS INDEX TO THE SIX TABLES.

THE CONTENTS OF THE FOLLOWING HISTORY OF THIS WORK.

INTRODUCTION.

THE *imperfection of the common methods of making anatomical figures, and by what steps the author was led to the true method; viz. making the skeleton the rule and foundation of his figures.—Eustachius used a like method.*

CHAP. I. Of the Skeleton and its Figures.

SECT. I. *Exact figures of the skeleton first to be procured.—These can only be had by clearing a skeleton of the soft parts, leaving only the natural ligaments, then placing it in a proper attitude, and fixing it so.—How this was done, by means of a tripod to support the pelvis below; and by stretching ropes to the ceiling and walls, from the head, trunk, and extremities, to fix the other parts; and afterwards comparing and correcting the whole, by a naked man placed in the same attitude.*

SECT. II. *The method contrived by the author, of drawing the general figures of the skeleton, thus placed and fixed, with the articulations; so as to be seen distinctly by the artist, and yet as at the distance of forty feet.—His precautions, and the accidents and difficulties that occurred.—Particular parts afterwards added, from the single bones when cleaned.*

SECT. III. *Reason for the attitudes that were chosen.—An exact and beautiful description of the attitudes, and of the position and bearing of each part and member, first of the front and back, and then of the lateral view of the skeleton, most useful both to anatomists and to painters.*

SECT. IV. *From what kind of body the skeleton was taken; the precautions used to procure a complete one, and even that corrected in the drawing.*

CHAP. II. Of the Muscles and their Figures.

SECT. I. *The author's method of inscribing the muscles on the linear figures of the skeleton, and the laborious preparations towards it.—Errors shunned, arising from the nature of reverses, of engraving, and the extension and contraction of the wetted paper.*

SECT. II. *The Muscles, and some other parts to which they refer, are annexed to the figures of the skeleton, as they are seen in dead bodies.—To make the muscles appear uniform, though taken from different bodies, they are referred to an outline of one muscular body, as well as to that of the skeleton, every thing being copied from nature.—To apply the muscles exactly, the exterior orders are first drawn, and afterwards corrected, as internal appearances required.*

SECT. III. *Great care taken, and various arts used, to exhibit the muscles in their true situation and appearances; as for example, in the diaphragm, the muscles about the anus, the velum of the palate, face, &c.*

Adult bodies were used, and these of the most perfect kind; things of less moment, and rare varieties are omitted, except a few; as the small psoas, and the muscle of the bladder.—The situation, figure, size, origin, cohesion, fleshy parts, and tendons, with the course of the muscular fibres, were chiefly attended to.

SECT. IV. *The muscular tables are of two kinds. The first exhibit, in as few figures as possible, the uniform system of the whole by successive orders, in front and back views, and in one lateral aspect, in proper attitudes, corresponding to the skeleton, of a commodious size, and properly shaded; and to complete the system, the orders on the neck, and under the head, are added in a side view; also these on the sole of the foot, and in the eye socket.*

The second kind explain, and serve as a supplement to the former, being the figures of the particular muscles, in a size double to the others, and done in a more simple manner, but these of the internal ear are of the natural size.—The composition and internal structure of muscles are omitted.

CHAP. III. The excellency of the tables. Of the linear figures, marks of reference, press work, and index.

SECT. I. *The merits of the artist Wandelaar, and the pains taken by him and the author, for ten years, to render the work every way complete. The excellencies of the tables, both as to anatomy and the painter's art, beautifully described, and how to look at them.*

SECT. II. *The advantages of the author's method of adding linear figures, and putting the marks of reference on them.—The figures of the single parts and muscles did not require this, tho' they are done with equal skill and attention.*

SECT. III. *The care taken even in inscribing the marks of reference, also in the choice of the paper, and in printing the tables.—The explication is done in the manner of an index.*

CONCLUSION.

The difficulty, labour, and expence of the work vindicated, from the usefulness and dignity of the subject.—The particular improvements to be found in it, are left to the discernment of the reader.

THE

HISTORY OF THIS WORK.

INTRODUCTION.

THAT the nature of these TABLES may be better understood, it is proper to explain the plan and method by which they were constructed: and I shall narrate, not only the things I approved, but also what I condemned, that we may be better able to judge in the conduct of works of this kind. First then, I began with the muscles; desiring accurately to represent in my figures, not only the single ones apart, but also the complete system of the whole. I began with the system. This I was obliged to divide into different orders, as the muscles lie one behind another: the first order containing these that lie immediately under the common integuments: the second, those which appear when the first are removed, and so on of the rest. I inclined, that not only the position of the members should be the same in each of these orders, but that they should be, in every respect rightly connected, and following each other; so that the whole series of the figures should plainly show every thing as it is in the body, in the posture and aspect I had chosen. With this design I set these orders before my artist for imitation, in the manner that most other anatomists had done, and he copied them with all the exactness he was able; but found this great difficulty, that in drawing the first orders, it was impossible to express these muscles and bones, that at that time only partly appeared, (and must be drawn as they then appeared, tho' they were seen more fully afterwards when the exterior muscles that covered them were removed) so as they might be continued without error in the succeeding figures; and being so continued, would be found placed on the body as nature showed they ought to be. For tho' afterwards, such amendments were made in the figures of the exterior orders, as the interior orders showed to be necessary, and in this way we had a set of tolerable figures, which evidently showed a skilful artist, yet they by no means pleased me, because the parts did not properly hang together, surely not in the manner I desired; nor were they expressed with sufficient roundness and precision: in a word, they were totally different from what I had conceived in my mind. This method I had made trial of in several parts, *viz.* the belly, the breast, the arm, the face; and I derived this advantage from it, that thereby I discovered the method to accomplish what I desired. For I had observed in the course of drawing, that as the first orders of muscles, pretty nearly resembled in general, the figure of the several members to which they belonged, they could be very well expressed by the artist, inscribing them as it were, upon that figure of the body which he had in his mind: the drawing imitating

every

every where the figure of the ſeveral members; but as the more he advanced to the interior parts, by removing the external muſcles, the more this figure of the members was loſt, the artiſt was not equally aſſiſted by this figure, in expreſſing the interior orders, and was obliged to accommodate all the ſucceeding orders to the firſt, as to a foundation, as therein this figure of the members generally appeared: now, this did not anſwer in the bones and ſyſtem of the ſkeleton, as it is of itſelf a peculiar compoſition, correſponding on the whole to the figure of the human body, but differing from it in many reſpects; yet the muſcles could never be truly expreſſed, unleſs the ſyſtem of the ſkeleton to which they are affixed, was firſt determined. Beſides, I was convinced, that all would be vague and uncertain, unleſs every part was repreſented according to meaſurement; if not with the utmoſt preciſion, at leaſt, with conſiderable care. But here I likeways foreſaw great difficulties: firſt in the exact meaſuring itſelf, and laying down the parts on the table according to that meaſure; and next, tho' this were done entirely to my wiſh, I foreſaw other difficulties hardly to be conquered. For it is evident, that the body and every member, muſt be expreſſed in the table, in the ſame attitude in which it was meaſured: now, tho' ſome members could be rightly placed for this purpoſe; as the head, the arms, the hands, the legs, the feet; yet others could not, as the trunk and the neck: therefore I foreſaw, that theſe members which could not be ſo placed, muſt either be repreſented in an aukward poſition, or I muſt leave it to the artiſt, to correct it by his ſkill in drawing: the firſt was intolerable, and the other I could never approve; becauſe there was great danger of miſtakes, or even a certainty of them. And if it was poſſible to put every part in a fit poſition, yet by raiſing and handling the muſcles, or on any other occaſion, if the firſt poſition was a little changed, it ſeemed ſcarce poſſible to replace things ſo, as exactly to correſpond to what had been already drawn; eſpecially as it was clear, that one and the ſame body would not ſuffice for the whole, and that different bodies would not exactly agree; nor, if they did, could they be placed ſo exactly in the ſame poſitions, that the parts could be drawn by meaſure, ſo as to fit and agree with the former. All theſe circumſtances therefore clearly ſhowed, that what I aimed at, was a thing of greater moment, than to be attained by the ordinary manner, even of the moſt approved anatomiſts; who did no more than copy in their figures, merely what was preſented to the eye, and ſo produced vague figures, in which no regard was had to order, nor exact meaſures, nor to any ſeries and connection. Such figures, tho' they may be good in ſome reſpects, and contain much, yea a great deal, yet they are defective in many things much to be deſired, and which are neceſſary to that moſt difficult point, adding perfection to a work. In a word, it was evident that a very different method muſt be purſued, and that my purpoſe was only to be gained, by fixing a certain rule, by which the figures muſt be conſtructed. And as human bodies, tho' they vary from each other in many things, yet in many things in general agree, there muſt be taken from the body itſelf a common foundation, on which to form the figures, and this muſt be the SKELETON, which ought to be the foundation of the figures, as it is of the body itſelf and of the muſcles, if we deſire to produce any thing certain, and anſwering to nature. The figures of the ſkeleton muſt therefore be firſt formed, and to theſe the muſcles muſt afterwards be referred: for the figures of the ſkeleton being firſt determined, as in moſt men of whatever ſtature and thickneſs, the muſcles adhere and are contiguous to the ſkeleton nearly in the ſame manner, and in the ſame places, it muſt follow, that the muſcles will correſpond to theſe figures of the ſkeleton, tho' they are taken from bodies the moſt different (provided they are not deformed) if they are painted on the figures with the ſame reference to the bones as they have in theſe bodies. I found likeways another uſe of this rule, that by making the figures of the muſcles in this manner, we could thereby find and know many of

them

them in the living animal, at leaſt the places of moſt of them. For firſt, by means of the figures of the ſkeleton, it is no difficult matter to know moſt of the bones in the live animal; and of theſe which do not ſo openly appear, we can at leaſt diſcover the places and poſition, or find rules for doing it; and theſe being once known, we have the means of conſidering and judging how the muſcles, as they appear in the figures, are placed on the bones of a living animal. And beſides, the figures of the ſkeleton, and the ſyſtem of the muſcles being once determined, the figures of the bowels may be referred thereto, and theſe too being fixed, we can determine theſe of the arteries, veins, nerves, and other parts. For as architects having laid a ſure foundation, raiſe the building upon it with all its parts, ſo we may on the ſkeleton, as a foundation, annex the muſcles, apply the bowels, and other parts, and afterwards conduct the nerves, arteries, and veins among them, and ſuperadd whatever more belongs to the fabric of our bodies: and therefore we will beſt imitate by art, this method pointed out by nature. And this firſt enabled me to diſcover, that a like plan was purſued by EUSTACHIUS in his tables.

CHAPTER I.

OF THE SKELETON AND ITS FIGURES.

SECT. I. BUT leaving theſe firſt rudiments, which can only ſerve to ſhow us in what manner we are to proceed, let us now ſolely conſider how we are to form theſe figures of the ſkeleton. In order to their being good, and a true foundation to the other parts, it is required that they ſhould be accurate, and truly exhibit the exact figure of the ſkeleton, as it is in the living animal, in the poſture which is choſen. Now that every part of the bones, which was to be repreſented in the figures might appear, it was neceſſary to clean them perfectly, and conſequently to ſeperate them one from another: but when once taken aſunder, it would have been no eaſy matter to rejoin them, nor could it be eaſily diſcerned whether they were rightly joined or not, if we did not compare them with the natural compoſition itſelf: and how was this to be found? and even if it could be found, it was eaſy to foreſee, that tho' in general they might approach to this compoſition, yet by no means with that accuracy I wiſhed for, and which was requiſite for the true application of the other parts. Beſides, the cartilaginous cruſts with which the bones are covered in the articulations, by exact cleaning, are either ſpoil'd or totally loſt, and therefore if they were join'd in that condition, the articulations would be imperfect: and the more that I was ſenſible of theſe defects, in the figures of the ſkeleton that are already extant, the more I deſired to find a remedy. After long uncertainty in what manner to proceed, it came into my mind, to prepare and clean a ſkeleton, in ſuch a manner, that nothing ſhould remain but the ligaments that bind the joints, and being ſo placed, and afterwards drawn, as I intended, to cut open and remove the ligaments, that what was at firſt covered by them might be added to the figure: and in this manner I thought I could beſt expreſs the true nature of the parts. But I foreſaw how laborious a taſk it would be, to prepare ſuch a ſkeleton, and how difficult to place it properly, after it was prepared; and it

was

was to be feared, as the painting such a skeleton must require a considerable time, lest by the drying of the cartilaginous crusts, and of the ligaments, and also by putrefaction, it might be greatly spoiled, and likeways become offensive: wherefore I thought it best, first to try the experiment in the seperate members of the skeleton: in some it succeeded very well, as in the hands, the feet, the joints of the thighs; in others it was much more difficult, as in the ribs and spine; but so, that there appeared hopes of overcoming the difficulties. Being therefore encouraged by the success I had found, about the end of the year MDCCXXV, having procured a proper subject, I prepared therefrom such a skeleton as I have mentioned; which being of itself unstable, as the ligaments were naturally soft and lax, I considered in what manner I could place and fix it in a proper attitude. To dry the ligaments, so as to stiffen the whole, was not proper, lest the composition of many bones should thereby suffer, and the cartilaginous crusts of the articulations be spoiled; and if it had been proper, it could not be done, till the skeleton was first fixed, in the position in which it was to be drawn. Therefore in order to place it, I took the following method. As the feet could not support the trunk, because they were neither rigid themselves, nor could the pelvis rest firmly upon the heads of the thigh bones, it was my first care, to support the lower part of the trunk upon a firm and stable basis; but in such a way, that it might be at liberty to incline a little on occasion, and in the manner that might be needful: for this purpose I ordered a tripod to be made, with the feet at a moderate distance; from the top of which, where the feet met together, a fulcrum of iron arose, which soon was divided into three branches, moderately seperated; of these one was shorter, the two others were equal in length; and all of them were in the upper part, to their extremity, first bent outwards the length of a cubit, in order to support, and afterterwards turned upwards, to retain whatever was placed thereon; the tripod being of such a height, that while the lower part of the trunk rested on it, the feet of the skeleton hanging down, could not quite reach the bottom of the tripod, and consequently the table on which it was placed. This tripod was placed upon a low table, that the lower parts of the skeleton, and especially the feet, might be more easily drawn; for had it been placed upon the ground, the artist, in order to see them aright, would have been obliged to stoop. The bottom of the trunk was so placed on the tripod, that the synchondrosis of the pubis rested on the extremity of the shorter branch, and the lower part of the ossa ilium, just before the sacrum, rested on the extremities of the longer branches; as I had taken care that the divarication of the branches, should be fitted for receiving and retaining these parts of the trunk; as also that the shorter branch, should be about so much shorter than the other two, as I judged the lower part of the synchondrosis of the pubis, ought to be below the lower parts of the ossa ilium, which last rested on the longer branches. This firm foundation being laid, my first care was to raise the trunk, with the neck and head, to an erect posture; and I began with the trunk, as it was firmer and more stable than the neck and head: for this purpose I fixed a cord below the neck, to the superior part of the spine, where it is more firm and stable, and I conducted it straight to the ceiling of the chamber, and having passed it through a ring fixed there, I conduct it to a hook in the neighbouring wall, round which I tye and fasten it. By the stretching of this cord I raised the trunk, as well as I was able, to an erect posture, but so that its lower part should still rest upon the tripod. I next pass another cord behind both the zygomatic processes, one end of it being under the right, and the other under the left, and bringing the middle of the rope to the occiput near the neck, I tied its two extremeties together like a handle above the head, and to this handle I fix another rope, which I likeways convey to the ceiling, and there passing it through another ring fixed near the former, I in like manner convey and fasten it to a hook in the wall. This rope I stretch as much

as

as I can, taking care at the ſame time not to relax that rope, which was faſtened to the ſpine; by this means the trunk was indeed erect, as alſo the neck and head, but not ſo completely, as to be quite free from inclination; for which reaſon tying ſeveral cords to the trunk, I conducted them in different directions to the walls, and faſtened them to hooks fixed there; and by ſtretching theſe cords, I fixed the trunk equally all around. Theſe laſt ropes, like the two former by which I erected the trunk and head, I faſtened to the more ſtable part of the trunk, viz. to the ſpine, being the foundation of the trunk, which the ribs muſt of courſe follow; and I tyed them to the ſuperior part of the ſpine, that I might thereby govern the whole of it, and I faſten them juſt below the neck, becauſe in that part the ſpine is ſtable; for had I tyed them to the neck itſelf, I ſhould have bent it in ſtretching the ropes, on account of its flexibility. Having in this manner fixed the trunk, I proceeded to the arms. And firſt I paſſed a rope, and fixed it to the conjunction of the clavicle with the ſuperior proceſs of the ſcapula; and by means of that rope, I was able to raiſe the whole ſhoulder to a proper height, and to ſuſpend the whole arm from the ceiling; and by tranſverſe ropes, I prevented the ſcapula from declining to the fore or back parts; and in this manner, having in general fixed both arms, I faſtened a rope to the inferior part of the right radius, and thereby I removed that whole arm from the trunk. Another rope I faſtened to the inferior part of the left arm bone, and thereby I raiſed that arm; another I faſtened to the left ulna, by which I governed the fore arm. The inferior extremities I fixed in the following manner; the right one I extended to a ſtraight line, and placed it directly under the trunk; and as it did not quite reach the table, I placed a piece of board between the heel and table, ſo fitted, that it fixed the heel as it were in a ſtanding poſture; neither was the limb ſo preſſed upwards, as to raiſe the pelvis from the tripod; and under the remaining part of the foot, I placed ſeveral boards, of ſuch thickneſs, that the whole might reſt in an equal manner. In the laſt place, I fixed a rope to the lower part of the thigh, and conducting it backwards to the wall, I thereby fixed the knee. Nearly in the ſame manner I alſo fixed the left limb, bending the knee a little, and raiſing the heel, ſo that the extremity of the foot reſted gently, on that part that is near the root of the great toe. The general poſition of the ſkeleton being thus fixed, I afterwards brought it to perfection; for partly by bending and raiſing the pelvis; partly by ſtretching or ſlackening the ropes, or by adding new ones; and partly at laſt by little boards, papers, cloths, or whatever elſe of the kind was readieſt, placed under, upon, or between certain parts, (the particulars of which is of no moment to relate), I corrected whatever I found defective. After this I placed a naked man, of a like ſtature and lean, in the ſame attitude, and with him compared the ſkeleton, eſpecially the pelvis, the ſpine, the thorax, the ſcapula and clavicles; for theſe being once truly placed, the other parts would give no great trouble. And correcting in the manner I have juſt now narrated, what ſeemed to require it; I afterwards conſidered the ſkeleton for a few days, and by a ſlight ſtretching or relaxing of the ropes, and other ways producing ſmall changes, I tryed if I could ſtill bring it nearer to perfection; after which I again compared it with the naked man, leſt by the rigid ſeverity of my care, I might have departed from nature.

Sect. II. The ſkeleton being placed according to my deſire, my next care was to imitate it exactly in a picture. Now to draw after it merely by the eye, as painters commonly do, would have rendered every thing vague and uncertain, and would by no means have anſwered my intention; for it was impoſſible but the artiſt muſt have committed miſtakes, and conſequently I ſhould not have had ſuch a figure of the ſkeleton as I wiſhed for, ſo that I might not only hope, but even be certain, that it would ſerve as a proper foundation on which to inſcribe the muſcles. To meaſure

the

the circumference of the whole in general, with the poſition, magnitude, and figure of each part, would have been an endleſs work; nor could it be done without the aid of ſome unerring rule. It would have been eaſy to confine the ſight by a wooden parallelogram, compoſed of four ſtraight ſides, at right angles, the ſpace comprehended by which, being made at leaſt equal to the ſurface of the ſkeleton, and the whole of it equally divided by ſtretched cords, into ſquare ſpaces equal to each other, this to be placed directly before the ſkeleton, and the tablet upon which the figure of the ſkeleton was to be drawn, being divided by lines, in the ſame manner as the ſquare was by cords, the artiſt looking through a ſmall determined hole, in the place from which he was to view the ſkeleton, would obſerve what parts of the ſkeleton were oppoſite to the different cords of the ſquare, and to what parts of theſe cords, and accordingly draw them on the correſponding lines of his tablet. But to this method there were objections, for in order that the artiſt might rightly and conveniently ſee every part of the ſkeleton, it was neceſſary that he ſhould not be at too great a diſtance from it; my inclination was, that he ſhould ſee it at a diſtance, not much leſs than forty Rhineland feet, as we call them, that thereby he might not ſee many of its parts obliquely; but as from this diſtance, the eye was not able to diſtinguiſh the ſmaller parts, therefore that the artiſt might be ſo near as to ſee the object diſtinctly, and yet might ſee every thing as at the diſtance of forty feet, the obſcurity only excepted, I contrived in the following manner: ſuch a ſquare as I have mentioned, and which I ſhall call the larger one, I placed immediately before the ſkeleton, ſo that the cords touched the moſt advanced parts of it. Four feet before this, I placed another ſquare, in every reſpect ſimilar, only the ſpaces were ſmaller; this therefore I call the leſſer one: the ſpaces were a tenth part ſmaller, as the diſtance of four feet, was alſo the tenth part of the whole diſtance I intended between the object and the eye. Theſe two ſquares I ſo placed, that the planes of the cords ſhould be parallel to each other, and in a perpendicular direction, and that the cords of the one, ſhould exactly correſpond to the other, the centre being placed directly oppoſite to the middle of the left part of the breaſt of the ſkeleton. This being done, the artiſt placing himſelf where he thought proper near the ſkeleton, at the moſt convenient diſtance for ſeeing it, contrived it ſo, that he ſhould plainly ſee ſome point of decuſſation of the cords of the ſmaller ſquare, exactly fall upon the correſponding point of the greater; and that part of the ſkeleton which appeared directly oppoſite to theſe points, he inſcribed upon the decuſſation of the correſponding lines of his tablet, which was divided by lines croſſing at right angles, as the ſpace of the greater ſquare was by cords; and in like manner paſſing from one point of decuſſation in the cords to another, the parts of the ſkeleton, if any ſeen directly behind them, were inſcribed on the correſponding points of decuſſation on the tablet; and the intermediate parts of the ſkeleton, between theſe determined points, were eaſily drawn without any error worthy of notice, on account of the ſmallneſs of the ſpaces. In this manner (ſo as to anſwer my purpoſe, tho' it gave the artiſt no ſmall trouble) was the fore part of the ſkeleton drawn as it ſtood, and it was drawn with the ligaments tying the joints, which being afterwards cut, and pulled back, as far as was neceſſary to ſee the joints of the bones, the artiſt added theſe parts to the figure. This being done, all the cords were looſed that fixed the ſkeleton, except the two drawn to the ceiling that kept it erect, viz. one from the head, and another from the upper part of the ſpine; and the ſkeleton with its tripod being turned, to exhibit the poſterior parts, it was fixed, and its figure drawn as the fore part; and I cut the ligaments of the joints in ſuch a manner, that tho' they were likewiſe cut upon the fore part, yet a ſufficient quantity ſtill remained to fix the joints, till the ſkeleton being again turned, the lateral view was likewiſe taken in the ſame manner.

But

But it being impoſſible to finiſh all the three figures of the ſkeleton, in leſs than nearly three months, tho' the utmoſt diligence was uſed, I was obliged during that time, to prevent the ſpoiling of the ſkeleton by drying, or by putrefaction, and alſo the inconvenience or miſchief that might thereby ariſe to ourſelves; ſome times therefore to prevent drying, whilſt I moiſtened with water, and poured it into the inciſions of the joints, to preſerve the cartilaginous cruſts; and ſometimes to check putrefaction, whilſt I ſprinkled with vinegar, and covered the parts during the night with papers and cloths moiſtened therewith, and poured vinegar into all the parts I could; it happened, that, while the firſt figure was a-doing, a ſmart froſt coming on, froze the whole ſkeleton, which was the moſt effectual thing, not only to fix its poſition, but alſo to prevent putrefaction; and if the froſt had laſted, till the firſt figure was finiſhed, I could eaſily have looſed the ligatures, and turned it, while yet rigid by the froſt, in order to draw the ſecond; but the thaw coming on ſooner than I wiſhed, haſtened the putrefaction, and gave me a good deal of trouble. The fire was likewiſe hurtful, without which the naked man neither could, nor would ſtand, till the weather became milder.

By theſe three figures, the poſture of the ſkeleton, the poſition of all the bones, and the compoſition of the whole bony fabrick was accurately expreſſed; but the figure and appearance of the bones only in general, for the remains of the ligaments about the joints, and other things ſtood in our way, from which the bones could not eaſily be cleared; nor would it have been proper, tho' I could have cleaned them in the moſt perfect manner, to have ſpent at preſent the length of time neceſſary thereto, eſpecially as I could do it afterwards, and have the figures completed at my leiſure. Therefore taking the ſkeleton to pieces, my next labor was accurately to clean each bone, ſo as to ſpoil nothing; yet when I had cleaned them, I delayed finiſhing the figures of the ſkeleton from them, till I had the figures of each bone, in the natural ſize, engraven upon copper; and I delayed it for this reaſon, becauſe as I intended to give attention to each bone, the making theſe figures was a preparation to the artiſt for that end, and on that account I could not return to the figures of the ſkeleton, till the year MDCCXXXIII. When I returned, my firſt care was to reduce them from their natural ſize, to what you ſee in theſe figures. Then the artiſt, conſidering each bone in the ſame poſition it held in the figures, ſupplied what was wanting in the figures, and amended what ſeemed to require it; after which, I was confirmed in the deſign I had always had, of publiſhing theſe figures, in caſe they anſwered my expectation; and I deſired they ſhould be engraven on copper, hoping that they would thereby be rendered ſtill more perfect; accordingly they were engraved, and from the bones themſelves, in order to a more complete expreſſion.

Sect. III. They exhibit the ſkeleton in an erect poſture, having regard likewiſe to the beauty of the attitude. I have choſen ſuch poſitions and attitudes of the members, by which the general arrangement of the bones and muſcles might beſt appear, and by which the viſcera, the arteries, veins, and nerves, and other parts might likewiſe appear, in caſe I ſhould incline to proceed. The poſition of the firſt and ſecond figure is exactly the ſame, repreſenting the fore and back parts, the one anſwering to the other, by which the continuation of the whole round of the figure may appear. A third figure is added giving a more complete lateral view. The poſition of this figure differs from the other two, being perhaps more proper to ſhow the lateral parts. Now as to the poſition, the following things are to be obſerved. The ſkeleton of the firſt and ſecond figure, ſtands firm upon the right foot, leaning only ſlightly upon the left; the right foot reſts upon the heel,

heel, and besides upon the anterior heads of all the metatarsal bones, especially that of the great toe, with the intervention however of the sesamoïdal bones; and the toes being bent downward, as it were lay hold of the ground, and by that means the foot stands firmer. The extremity of the foot is turned outwards, in a natural manner, in that position that gives the greatest firmness when we stand. The right knee is extended to a straight line, the patella lies upon the thigh bone, higher than that sinus between the condyles, upon which it commonly rests; and it lies as when the knee is straight and fixed by the action of the rectus, vasti, and crureus muscles. The leg is inclined a little outwards upon the foot, by the bending of the joint of the talus on the heel bone, by which the extremity of the foot rests fully upon the ground; and the heel is in the same perpendicular line with the head, which makes a firm standing posture. And this is assisted by a moderate inclination of the thigh to the same side, the thigh bone meeting the tibia so as to make an angle, but a very obtuse one, and making it towards the side I have mentioned. The left foot is somewhat removed from the right, both to aside and forward, and it rests only, and that slightly, upon the anterior head of the metatarsal bone of the great toe, by the intervention of the sesamoïdal bones. The left knee is moderately bent, and thereby the patella rests upon its sinus between the condyles, therefore the right extremity alone upon its vertex sustains the pelvis, whose position is oblique, the left side being the lowest, because as I said, the left foot is removed from the right, and yet touches the ground; and because the left foot is likeways advanced forward, the left part of the pelvis is for that reason forced somewhat more forward than the right, and the rest of the trunk above the pelvis is inclined towards the right, as much as is required for equilibrium. Therefore the whole spine in general is bent towards the right, and it is likeways, except in the neck, inclined as it were somewhat to the left; that while the left part of the pelvis, is advanced somewhat more forward than the right, the breast may yet stand directly forward. On the other hand the neck is bent towards the right, and the atlas with the head turned to the same side, as much as is necessary to direct the face somewhat that way. Besides, the pelvis is so situated, that its whole superior edge is placed obliquely, chiefly directed upwards, but at the same time remarkably forwards. Therefore the os sacrum, descending from the loins, inclines somewhat backwards, from which the coccyx advances forwards, at the same time bending itself that way. The loins rising from the os sacrum, are at first remarkably bent, then becoming more straight, are inclined moderately backwards; and and being concave behind, thereby better support the thorax. From thence, the dorsal part of the spine inclining likeways backwards, is gently bent from the top, but in a contrary way to that of the loins; by which the thorax does not incline too much forwards. From the back, the neck raises itself forwards, moderately bent, sustaining the head in such a manner, that the face may be thrown sufficiently forward. The thorax is moderately bent towards the right, along with the spine, by which the ribs of the right side approach nearer to each other; the superior ones being drawn a little downwards, the inferior upwards. On the contrary, the left ribs are more separated, the superior being drawn upwards, the inferior downwards. And for this reason the external circumference of the thorax on the left side is from above downwards wholly convex, but on the right side below the middle, is moderately concave; and therefore the lower ribs of the right side, being more erect upon the spine, than those of the left, their anterior extremities are more distant from it. The right arm is in a hanging posture but moderately raised; whereby the scapula is upright, and the clavicle almost directly transverse. But the scapula is moderately thrust backwards, and along therewith, that part of the clavicle that supports it, by which the breast is more open, the left arm is raised higher, and therefore that part of the clavicle, upon which the superior process of the scapula rests, is raised along therewith,

and

and the ſcapula undergoes ſome rotation, ſo that its lower angle is turned towards the left ſide. The right fore-arm is ſtraight, and ſo likewiſe are the radius and ulna; and the right hand hangs open. But the left fore-arm is ſomewhat bent, and the radius is turned, as much as it is capable of, round the ulna, and along with it the hand. So much for the poſture of the firſt and ſecond figures.

In the third table, the ſkeleton is placed as it were in a walking poſture. Like the others it ſtands alſo on the right foot, the left only reſting on the ground, upon the end of the great toe, which is therefore a little bent upwards, as in walking, when we are juſt going to bring forward the foot that is behind. The right foot ſtands upon the heel, and the anterior extremity of the metatarſal bone of the great toe, reſting on the interveening ſeſamoïdal bones, upon theſe it ſtands chiefly, and alſo upon the anterior extremity of the metatarſal bones of the ſmall toes. The right knee is ſtraight as in the former figures, and in like manner from the ſame cauſe, the patella is drawn upwards, ſo that it lies upon the ſinus between the condyles of the thigh, only by its inferior part, its upper part reſts upon the thigh above that ſinus. The left knee is moderately bended, therefore the ſuperior part of the patella reſts upon the ſinus, between the condyles; from whence its extremity is directed towards the eminence of the tibia, to which the ligament is affixed that proceeds from that extremity. The pelvis, as in the firſt and ſecond figures, reſts upon the right foot alone, and the left ſide is the loweſt; the ſpine above it is ſomewhat inclined to the right, whereby the fore part of the thorax is a little inclined that way; but the face ſtill more, not only becauſe the neck is more twiſted, but chiefly becauſe the atlas and head are remarkably turned that way. The pelvis is likewiſe ſo placed, that its ſuperior edge is directed upwards and forwards, as in the former figures, and the coccyx with the inferior part of the ſacrum are bent forwards. The loins near the os ſacrum, are at firſt conſiderably bent, then becoming ſtraighter, they incline moderately backwards, being concave behind; from theſe the dorſal part of the ſpine inclines backwards, and is gently bent from the top, concave before; from thence the neck ſtretches forward, moderately bended. As the left arm is raiſed, the ſcapula is ſomewhat turned, ſo that the upper part is directed a little backwards, and the inferior angle forwards. But as the right arm hangs down, and at the ſame time is carried backwards, the baſe of the ſcapula on this ſide is ſomewhat removed from the ribs, chiefly at the lower angle. This is ſufficient with regard to the poſition, what remains on that ſubject, in this or the former figures, may be eaſily perceived from the figures themſelves. All the joints appear full, and exactly fitted, becauſe the cartilaginous cruſts are not neglected.

Sect. IV. I muſt now deſcribe the nature of the ſkeleton, which I made choice of for this repreſentation. As to the age, I chooſed that, where the bones had come to their full growth and perfection, that is when the epiphyſes were plainly continued to their reſpective bones; for before that time, the bones are incomplete. I made choice of the male ſex, and of a middle ſtature, where at the ſame time all the bones had their juſt proportions and ſymmetry; and I chooſed ſuch a ſubject, as I looked upon to be more perfect than common, and which had nothing faulty, either in the bones themſelves, or in their compoſition. But as ſkeletons differ from each other, not only in age, and ſex, and ſtature, and in the perfection of the bones, but likewiſe in the marks of ſtrength, and in the whole habit and appearance, I chooſed one that expreſſed both manly ſtrength and agility, where all the parts diſtinctly appeared, and that in a moderate degree, ſo as neither to have a youthful and effeminate ſlenderneſs and ſoftneſs, nor too much roughneſs and want of poliſh; in a word, ſuch as had a beautiful and graceful appearance. For I wiſhed to take my example of nature, from nature in all her perfection. But as even the beſt ſkeletons differ one from another, and I wanted to exhibit only a particular one, this I pitched upon as an example for the reſt. And I confeſs I was happy in finding a body, which as it promiſed to contain, ſo it ſupplied me with ſuch a ſkeleton as I wiſhed for; but it was not ſo entirely complete, but that ſome imperfections occured; as painters therefore, when they draw after a fine face, and finding ſome defect, endeavour by ſkilfully mending it, to render the figure ſtill more beautiful, ſo I amended in the figures ſuch things as could not be approved; but I did it from the moſt approved originals, taking care never to depart from the truth. Such is the plan and hiſtory of the figures of the ſkeleton. From which I think it will clearly appear, that in this way the truth of nature may be correctly expreſſed, but that no ſkeleton could be actually preſented to view, as theſe figures repreſent, at leaſt it would be no eaſy matter. For who could undertake, after perfectly cleaning the bones, and conſequently removing the ligaments, that not only connected, but in part covered them, preſerving at the ſame time the cartilaginous cruſts, I ſay who could undertake after this, to exhibit them, neatly joined together in every part, according to the truth of nature.

CHAPTER II.

OF THE MUSCLES AND THEIR FIGURES.

SECT. I. NOW theſe figures being drawn, I had more courage to inſcribe the muſcles upon them, after they were engraved on copper; and both from my love of the work in which I was engaged, and my deſire to improve anatomy (as far as I was able) I reſolved to try, if I could accompliſh what I had conceived in my mind. Nor did I attempt it altogether unprepared. For as the number of muſcles ſpread over the whole body is very great, as they run into each other, and as many difficulties perpetually occur to one that deſires to underſtand them, not in a careleſs, but in a full and accurate manner, as there are many and great diverſities among them in different bodies, and as few ſubjects are fit to ſhew us nature in her greateſt perfection, and as it was convenient to avoid delays in the drawing; for theſe reaſons, it was neceſſary to provide before hand, as far as we were able. Wherefore from the time that I began to prepare the firſt figures of the ſkeleton, every year, while I was diſſecting the muſcles for my pupils, and likewiſe on other occaſions, I obſerved their poſition, connexions, their figures, thickneſs, and their parts, and either to confirm what I had remarked, or to add the variations that different bodies afforded; and this I continued to do every year. And becauſe things are better known from nature than from deſcriptions, and can thereby be preſented to the eyes, I kept ſuch as could be preſerved, that they might be of uſe on future occaſions; but eſpecially bones, and other parts in which muſcles are inſerted, with their extremities, by which they are ſaid to ariſe or be inſerted; theſe choſen from the moſt perfect bodies, I preſerved in a proper liquor, ſo that they ſuffered no injury

injury, and could be examined on every occafion, and as I now was poffeffed of many fcattered obfervations, I arranged them fo as to be fit for ufe. This being done, I compofed from thefe materials a hiftory of the mufcles, choofing what I found moft frequently to occur, and what I thought moft confonant to the intention of nature. And tho' it was my inclination to infert nothing in this hiftory, but what I found in the book of nature, yet I thought it would be ufeful to confult likewife the books of anatomifts, not only the moft reputable ones, but alfo fuch others as I could procure, that I might know, if I had omitted any thing worthy of memory obferved by them; nor did I defift to add to this body of myology, tho' publifhed, what further diffections afforded.

Thus prepared, in the year MDCCXXXVIII, I began to infcribe the mufcles upon the figures of the fkeleton, with more certain knowledge, and greater hopes of fuccefs. For this purpofe I ufed the outline figures of the fkeleton, the fhaded ones being lefs proper, by obfcuring what was drawn upon them. And here I took care to fhun an error, which I had obferved in making the tables of the fkeleton. When engravers transfer a drawing from paper upon the copper-plate, they firft rub over the back fide of the paper with cerufs, next they fkilfully lay that fide upon the plate, fitting it fo as the pofition of the figure requires, and then fix it there; afterwards with a needle flightly, but with a fit degree of preffure, tracing the lines of the figure, they find when the paper is removed, that thefe lines are marked upon the plate by means of the cerufs. The figure being engraved, and printed off, what is on the right hand in the plate, is found on the left in the paper printed from it, and *vice verfa*. If therefore the feries of the mufcles is infcribed on thefe printed papers, which afterwards by the engraver are applied to the copper-plate, as I have juft now defcribed, in order for engraving, in that cafe the right and left of the feries of the mufcles, would not anfwer to that of the fkeleton. This inconvenience might eafily have been fhunned, if from an impreffion taken upon paper, which I fhall call the architype, or firft impreffion, a fecond was immediately taken upon another paper, which would produce a contrary impreffion, as to right and left, to the firft, and therefore this might be ufed in order to incribe the mufcles (I fhall call it the antitype or reverfe), and then to be placed upon the copper, and the figure traced and engraved from it upon the plate; hence the impreffions from the plate will be contrary to the antitype or reverfe, and confequently agree with the architype, which was the thing required. But here the following difficulty occured; in order to have good impreffions, it is neceffary that the paper be firft foftened by maceration; and in this condition the two papers, the architype and the reverfe, being expofed to the rolling prefs, are extended to a larger fize, and confequently the figures upon them rendered larger, nor by drying are they contracted to their former dimenfions, therefore fuch antitypes or reverfes of the fkeleton being ufed for infcribing the mufcles, would have been larger than the figures of the fkeleton. In order to prevent this, I ufed dry paper, in taking the impreffions both of the architypes and reverfes, by which indeed the impreffions were bad, efpecially that of the reverfes, but yet they fufficiently anfwered the purpofe. But before I obferved this defect of the moiftened reverfes, the mufcles were already infcribed upon fuch reverfe of the fecond fkeleton; from which it happens, that the outline figures, which reprefent the pofterior order of the mufcles, are fomewhat too large. This I could overlook in thefe, being only intended to explain by their fimplicity and marks of reference the fhaded or finifhed figures, but I corrected it in the fhaded figures themfelves before they were engraved, for the fhaded figures were all engraved pofterior to the linear ones, (each linear figure being transferred to the plates, upon which their correfponding fhaded ones were to be engraved), and I corrected the error in the following manner. . I ordered dry

paper

paper to be used, in casting off the architypes or first impressions of the linear figures, that represent the posterior orders of the muscular system, but wet paper for the antitypes or reverses, by which the former were not enlarged by the force of the press, and therefore the size, and likewise the size of the reverse, when first taken from it, was the same as that of the figure upon the copper, from which the architype or first impression was taken. But as the paper of the reverses was wet, it contracted to a smaller size in drying, and consequently the figures printed on it; and thus by frequent trials, I so ordered the maceration of the paper, that in drying, it was just so much diminished, as was required to make the figure of the proper size; and I used the reverses, corrected in this manner, for transferring upon the copper the posterior orders of the muscular system, in order to engrave the shaded figures. From this however I perceived, that when the finished tables came to be published, tho' they were all of the same size upon the copper, yet it could hardly be prevented, but the different copies must somewhat vary in their size, according as the paper on which they are printed, happens to be more lax or firm, or as the paper was more or less macerated; for according to these circumstances, the contraction in drying will be more or less.

Sect. II. Now the muscles in the drawing were referred to these linear reverses, in the same manner as they are situated and fixed to the skeleton in a dead body; and the other members to which muscles refer, as well as to the skeleton, such as the os hyodïes, the larynx, the tongue, and others, were themselves in like manner referred to the skeleton, and the muscles to them. I at the same time consulted the history of the muscles, and the annotations added to it after publication, and also from time to time the preparations I formerly mentioned, yet every thing was taken from real bodies; and these things which manifestly and remarkably differed from the more frequent appearances, were also supplied from other bodies; now tho' this at first succeeded to my wish, but with considerable trouble, yet a new difficulty occurred. I could not possibly take the figures of the greatest part of the muscles, much less the whole, from one body; yea it was clear that some years must be employed, and many bodies used, and tho' no doubt the muscles could from any body be referred to the figures of the skeleton, yet as some bodies are much fuller than others, for I could not expect to find them all nearly alike, it was difficult to contrive how, taking and arranging the muscles from bodies of different fulness, I could reduce the whole to an equality and proportion one to another, for the skeleton alone was not sufficient for this purpose; therefore besides this, some other fixed thing must be found, for to trust merely to opinion, was neither safe nor pleased me; now this was chiefly wanted for the great thick muscles, especially for these situated upon the trunk and neck, and above all upon the extremities. I therefore took care from the first body to have an exact drawing of the exterior circumference of the trunk, neck, and extremities, as it is made by the muscles, within which they were arranged, and also referred to the skeleton, from whatever kind of body they were taken, yet in drawing these muscles, I still used bodies as like as possible to one another. But it being difficult, in drawing the exterior order of muscles to discover what position they have with regard to the skeleton, because it is almost totally covered by them; therefore, tho' the greatest care was taken, we were afterwards obliged to correct the exterior orders, according as the interior ones directed us, when the skeleton was more exposed to view. To apply the muscles accurately to the figures of the skeleton, each muscle ought to be separately applied, but in this way it would have been no easy matter to arrange them properly into a system. In order to give this arangement, we must begin with the exterior muscles, and in this way, as has been said, it was difficult to place the exterior ones aright in regard to the skeleton, it being totally covered by them. It was likewise difficult to place these muscles that were almost totally covered by others;

both

both difficulties were overcome by beginning with the exterior order, and proceeding gradually to the moſt interior, afterwards always correcting the exterior as the more inward pointed out.

In diſſecting the muſcles, and expoſing them to the painter's view, care was taken to hurt nothing. In order to this, when it was needful, I uſed bodies that were rather fat, the fat ſupporting the muſcles, and I took no more of it away, than was neceſſary to ſhow what I wanted to expreſs, leaving the reſt untouched, in order to ſuſtain. It was neceſſary to diſſect many muſcles, and to draw them by parts, eſpecially theſe which would have been ſpoiled, or in great danger of it, had I laid them bare and expoſed the whole at once; and on occaſion I was obliged to call various methods to my aid; thus, in order to give the figure of the caracohyoïdeus, which is the 35th of the XIth table, its exterior part was firſt laid bare, and its figure taken, next the interior part of its origin; to exhibit which I took away the fat with the greateſt care, without hurting any thing, leaving ſo much as neither ſtood in our way, nor allowed that origin to ſink. The concave and convex parts of the diaphragm, as they are expreſſed in the IVth and XIVth tables, cannot be ſeen at one and the ſame time in the body; for to ſhow the concave part, the abdominal viſcera, which conceal it, muſt be removed; and to ſhow the convex part, the thorax muſt be opened. But when both theſe cavities are open at the ſame time, the diaphragm is relaxed, and exhibits a falſe appearance, both above and below; therefore removing the abdominal viſcera, I firſt expoſed the concave part, and after the figure of it was taken, I replaced the viſcera, to ſupport it in its true poſition; then opening the thorax, I added the convex ſide of the figure; and in order to render it complete, I opened the thorax of another body, the abdomen being entire, and its viſcera ſupporting the diaphragm. Many contrivances were neceſſary to procure a proper view and figures of the muſcles of the anus, and eſpecially of the pharynx, the ſoft part of the palate, and the face; by which, if I am not miſtaken, many things are truly repreſented, of which ſome can with difficulty be fully ſeen in the body, others by no means without theſe contrivances. But it would be too tedious to narrate the methods I uſed, to ſhun every falſe repreſentation.

Adult bodies were made uſe of, and of theſe ſuch as ſeemed moſt proper for our purpoſe, and the muſcles were exhibited according to their moſt frequent appearances, and ſuch were choſen, that we had reaſon to believe were moſt perfect and preferable. It would have been endleſs to purſue all the varieties, that are obſerved in them, as in the outlines of the body. And even paſſing theſe ſlighter ones, had I been diſpoſed to purſue the more ſtriking and remarkable varieties, I ſhould have found it a very tedious affair; nor was it proper, in this univerſal map or plan of the muſcles, to inſert many varieties, even of theſe that were frequent and remarkable. Yet ſome are inſerted, even ſome few that rarely occur, as the ſmall pſoas, and the muſcle of the bladder, which I have very rarely ſeen, and altho' I took the utmoſt pains, to diſſect and paint them all, in the moſt full, accurate, and ſubtile manner, yet I omitted certain things of leſs moment; thus ſome that are a little tendinous at both extremities, are not ſo in the figures, and certain fiſſures are omitted, thro' which ſmaller arteries, veins, and nerves paſs; and other things of the like kind, becauſe they ſeemed of ſmall moment, or would have rendered the figures and general courſe of the muſcles obſcure, at leaſt would have ſpoiled that ſimplicity which I aimed at. For it is ſurely proper to uſe moderation, and a certain judgment in theſe matters, as reaſon and the nature of the thing require. The ſituation, the figure, the magnitude, the origin, the inſertion, the coheſion, the fleſhy and tendinous nature, the general courſe of the fibres, to which their direction may be referred, were the chief things I aimed at in theſe figures.

Sect. IV. I comprehended the whole relating to the muſcles in two kinds of tables: the one contains the ſeries of the muſcles over the whole body; the other contains the figures of the particular muſcles. The ſeries contains front, back, and ſide views, as in theſe of the ſkeleton; and the muſcles are repreſented by orders; firſt, the exterior, and afterwards the more interior, one after another, and every ſubſequent figure is a continuation of the preceding one. And becauſe the fore and back parts of the body are of greateſt extent, and fuller at leaſt in general, and as by comparing theſe two, we may in a great meaſure judge of the ſide view, for theſe reaſons the ſeries of the ſeveral orders have been exhibited from before and behind. But we have added a ſide view of the exterior muſcles, that theſe might appear more fully, than by a mere compariſon of the fore and back parts; this the poſition of the muſcles there ſeemed to require. And I thought this firſt ſide view was ſufficient, as the general nature of the ſeries was eaſily ſeen from thence, eſpecially if the fore and back views be compared, and likewiſe when needful the figures of particular muſcles. But certain orders of the muſcles, ſituated on the neck, and under the head, are repreſented in a ſide view, as they could not be expreſſed in a fore or back view, or indeed in any other, ſo well as in a lateral one; and for the ſame reaſon the orders of muſcles on the ſole of the foot are repreſented, and alſo theſe in the eye-ſocket. Now, tho' according to the varieties of poſture and ſituation of the members, as well as points of view, the ſyſtem of the muſcles ſhows itſelf in an infinity of different appearances; ſo that figures might have been multiplied without end; yet we have made choice of the moſt convenient poſition, and as there is no poſition in which ſome part or other does not imperfectly appear, I have choſen that, which ſhows the ſyſtem in general in the beſt manner; and even in this way I could have made many more orders, but as the few which I exhibit ſuffice, in my opinion, for giving a general view of the whole ſyſtem, and if any thing be wanting it may be eaſily ſupplied by comparing the orders with the particular muſcles, for that reaſon I kept within certain bounds; and was at pains to limit the number, leſt a multitude ſhould produce confuſion. And I found it a more difficult task thus to repreſent the whole ſyſtem with propriety, in a few orders, than merely to have multiplied the orders. But as none of the muſcles, except a few, can wholly appear in theſe orders, it became neceſſary to add complete figures of the particular muſcles; and even tho' I had ſo multiplied the orders, that every particular muſcle would have been ſomewhere ſeen totally naked, yet all of them would not have been ſo rightly known, at leſt not ſo eaſily and readily, from theſe, as from the particular figures, which are in no reſpect diſturbed or obſcured by neighbouring muſcles. Beſides, the figures that repreſent the ſyſtem muſt not be too large, that they may be under the view at once, and eaſily handled; ſo that they could ſcarce be larger than they are in theſe tables, and even theſe may perhaps be thought too large. However the ſize is a proper one, for expreſſing moſt, even of the ſmall muſcles, in ſo far as to ſhow their connexion as a ſyſtem; and it would alſo have been a ſufficient ſize, to give a fuller knowledge of the larger muſcles by particular figures, and even for many of the ſmaller; but in others, not a few, either from their ſmallneſs, or the manner of their compoſition, it would not have anſwered; beſides, the nature of the light and ſhades could not, in the ſyntactic figures, be ſo well expreſſed by that ſimplicity of lines, which is beſt fitted to expreſs the courſe of the fibres, (and which for that reaſon we have choſen in the figures of the particular muſcles) as they are by the decuſſated manner there uſed. It was neceſſary therefore, to reduce the ſyntactic figures to as few orders as I was able, and to finiſh them in their own manner, and beſides to exhibit the particular muſcles in ſeparate figures.

In

In the figures of the particular muſcles, I have followed the ſyntactic ones, wherever I was able, to the end that the former might ſerve to illuſtrate the latter, and to ſhow more fully and clearly, what in theſe laſt was, by reaſon of incumbent or adjacent parts, hid and obſcure, or could not there be ſo well expreſſed; beſides, in this manner, every thing was more coherent. But when the poſition taken from the ſyntactic figures was not ſufficient, I added another more convenient poſition of the ſame part. Some muſcles, whoſe poſition was improper in the ſyntactic figures, and ſome few, which could not at all be ſeen there, I exhibited in the poſition that ſeemed moſt proper. I could have greatly multiplied the figures, had I inclined to ſhew the ſame things in every poſition, as they appeared on the exterior, interior, and lateral parts; but I rather chooſed to proceed with a certain judgment, and to exhibit only ſuch views as were ſufficient to the intention of this work. The figures of the particular parts are double the ſize of the ſyntactic ones, that they might be the fitter to expreſs every thing, but eſpecially the ſmall parts, in a fuller and more perſpicuous manner; and tho' the great muſcles did not demand this, yet for the ſake of uniformity, the ſame proportion is there retained. The muſcular parts of the internal ear, as they are called, being themſelves ſmall, are expreſſed in their natural ſize; and the figures are all of intire parts, unleſs that ſome detruncated were added, from the neceſſity of expreſſing certain remarkable things. But as to their compoſition, and internal ſtructure, they are omitted, not to ſwell too much the ſize of this work.

CHAPTER III.

THE EXCELLENCY OF THE TABLES. OF THE LINEAR FIGURES, MARKS OF REFERENCE, PRESS WORK, AND INDEX.

Sect. I. MOREOVER, I not only ſtudied the accuracy of the figures, but likewiſe their perſpicuity, and their beauty. Therefore I employed an artiſt, that excelled both in drawing and engraving things of this kind, and who (which is very rare) had a remarkable paſſion for works of anatomy, and who was confirmed therein, by my never refuſing him the price he demanded. This artiſt, who for many years paſt, devoted his work to few beſides myſelf, and for the laſt ten years (during which, except ſome little intervals, he was wholly employed on theſe tables) almoſt to me alone; and he drew and engraved every thing under my conduct, and I laboured, from time to time, that he might as much as poſſible underſtand the things he was to expreſs. I was afterwards preſent while he made the drawings, directing him how every thing was to be drawn, aſſiſting him, and correcting what he had drawn: and he was form'd, conducted, and even governed by me, as if I myſelf, by his hand, had drawn the figures; and afterwards when he came to engrave, much care was required, that he might commit no error in imitating the figures on the copper; and we frequently conſulted, what was the

beſt

beſt manner to engrave each particular. And as even with theſe precautions, errors were inevitable, I reviewed the figures after they were engraved, and he expunged the erroneous parts I pointed out, and reſtored them according to the truth. The principal care was to expreſs every thing truly, and in the cleareſt manner; but the artiſt likewiſe exerted his skill, not only in the outlines, and the light and ſhade, but likewiſe in the ſymmetry and proportion, and in the particular appearances of every part. He aimed at dignity in the outlines, clearneſs and force and grace in the light and ſhades, and likewiſe a proper harmony, ſo that every thing ſhould be fully ſeen, and at the ſame time the whole figure, tho' conſiſting of many united parts, ſhould ſeem no where interrupted, as far as the nature of the thing could bear. In the ſymmetry he ſtudied a certain congruity and equality, ſo that all the parts agreed one to another; in the particular appearance of the parts, that diſtinction and diſſimilitude, that bone, fleſh, tendon, cartilage, and other parts have to each other; and in the whole figures he ſtudied even a certain pleaſing appearance. To the ſyntactic figures he added back grounds, not only to fill up the blank of ſo large a table, and that the appearance might be milder, but alſo, by means of the temperament of the light and ſhades of theſe back grounds, that the light and ſhades of the figures themſelves might be preſerved; ſo that they might ſeem to riſe and ſtand out from the tables, and thereby alſo, tho' the figures are as it were broken, by conſiſting of ſo many parts, yet they appear ſolid and entire. And this was a thing that required no ſmall art in all the tables, but chiefly in theſe of the skeleton; the effect whereof will be beſt perceived by viewing the tables at a proper diſtance, applying the hand to the eye, in ſuch a manner as to prevent diſturbance from the ſurrounding light; nor do I imagine that the back grounds can hinder any one that uſes the hand, and is not a very careleſs obſerver, from readily perceiving whatever is repreſented in the tables.

Sect. II. Not to diminiſh theſe excellencies of art, we abſtained from inſcribing the marks referred to in the explication; becauſe they would not only have appeared ſo many blots upon the figures, but would have rendered many things obſcure, and would have even entirely obliterated not a few; as there are many parts ſo ſmall, that theſe marks would either have entirely filled them, or nearly ſo; and the marks themſelves, when placed in the ſhade, would either have been obſcure, or even quite inviſible. To all this a remedy was found, by adding the lineary figures, and inſcribing the marks upon them; from which we have alſo this advantage, that the extent and limits of every thing are readily and diſtinctly ſeen, in theſe lineary figures; whereby all doubt is removed, cauſed ſometimes by the ſmallneſs of parts, the nature of the ſhades, or of the engraving itſelf in the ſhaded figures. But the marks are inſcribed upon the figures themſelves of the particular muſcles, for as they chiefly repreſent ſingle muſcles, and in a larger ſize, and are engraved by one ſtroke, in a ſimple manner, theſe marks can eaſily be inſcribed upon them, and can as eaſily be ſeen and found; nor could they at any rate ſo much hurt theſe figures, where ſo many aſſiſtances of art were not required. As to the bones, and other things exhibited with the muſcles, in order to ſhow the parts they touch, or are connected with, theſe are only expreſſed by outlines; not only becauſe it was ſufficient, but alſo as in this way the outline, extent, and limits of the muſcles, more clearly appear. However, the muſcles themſelves are expreſſed with no leſs ſkill in theſe figures, whether you regard the exactneſs of the outline, or the light and ſhades, or the diſtinction of the tendinous and fleſhy parts. There is likewiſe a different maner of engraving, than in the ſyntactic figures, advantageous in expreſſing the courſe of the fibres by ſimple lines inſtead of decuſſation. This ſimple manner was preferred, in order more clearly to expreſs the courſe of the fibres, which is only expreſſed in a general manner, as I did not chuſe to repreſent the

the fibres, and the nature of muscular composition in a curious manner; for besides the impossibility of giving a true representation of it, I thought a general idea of the course of the fibres was sufficient, in this system of general figures.

Sect. III. To engrave the marks, I made use of a skilful engraver, who could execute it with judgment, by inscribing them exactly upon their true places, and by proportioning the size and fulness of the mark to the nature of the part; by which the parts were accurately pointed out, the marks themselves were conspicuous, and did no harm, especially by not obscuring small parts. I afterwards took care, that the tables should be printed off in the best manner, a thing (as artists well know) of great moment, not only for elegant neatness, but to express the full force and gracefulness of art. Therefore I made use of the fittest paper, whereon much depends, and also I employed an intelligent and experienced printer. As to the explications, I thought short ones in the manner of an index might suffice, but these belonging to the single figures are somewhat more full: for the rest I refer to my history of the muscles.

CONCLUSION.

THE above I thought proper to say relating to the nature of this work. But that person will understand it best, and will feel the difficulty of the undertaking, who shall heartily engage in a work of the like kind. It may be thought I should have treated, in a particular manner, of the advantages I have attained, and can promise myself, above the works of these excellent and praiseworthy men, who have gone before me in this road, by all these efforts, this labour and expence I have bestowed, which I must confess have been greater than any one would imagine. But such as love and cultivate these studies, by considering what I have said of the nature of the work, will easily see, what I would wish to have done for their advantage. And if any are desirous to know, exclusive of the general plan I have followed, wherein I differ, in real things, in my figures and writing, from these of former anatomists, and what things are either amended, or added; as such person may be satisfied by comparison, I thought I might be silent on that subject; especially as the labour I must employ on it, may be better bestowed, as it must needs be very great, in so great variety of things. But if any should be of opinion, that it is superfluous, with so great efforts, to seek after such accuracy and perfection, in a thing of no necessary use, let such people consider, besides the necessary utility, what the greatness and dignity of the thing itself requires; and then, I dare say, their wonder will be less, that I esteem these tables the more worthy of praise, not only on account of their fidelity and truth, but also in proportion as they are more excellent and perfect.

Leiden, MDCCXLVII.

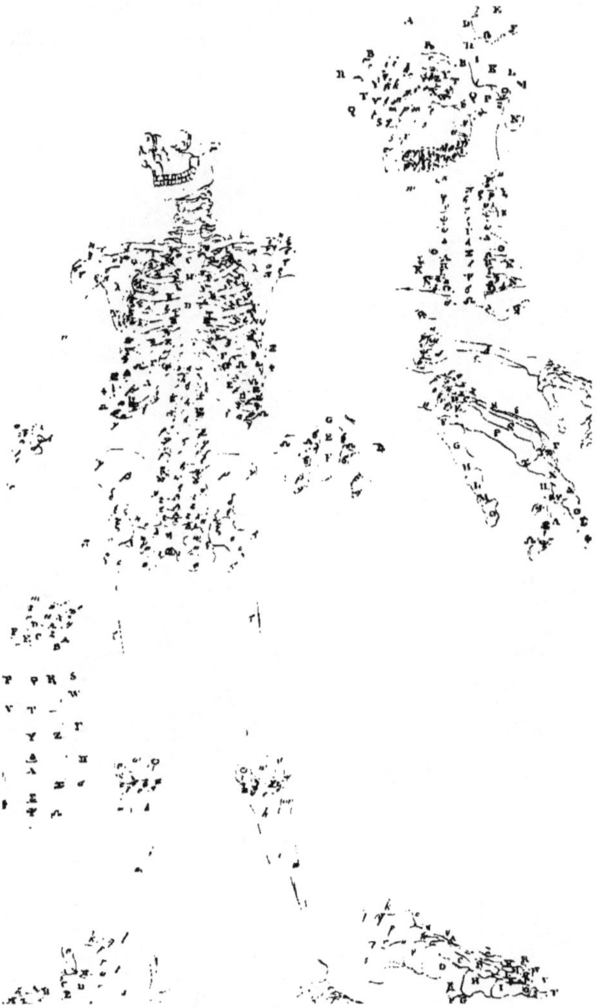

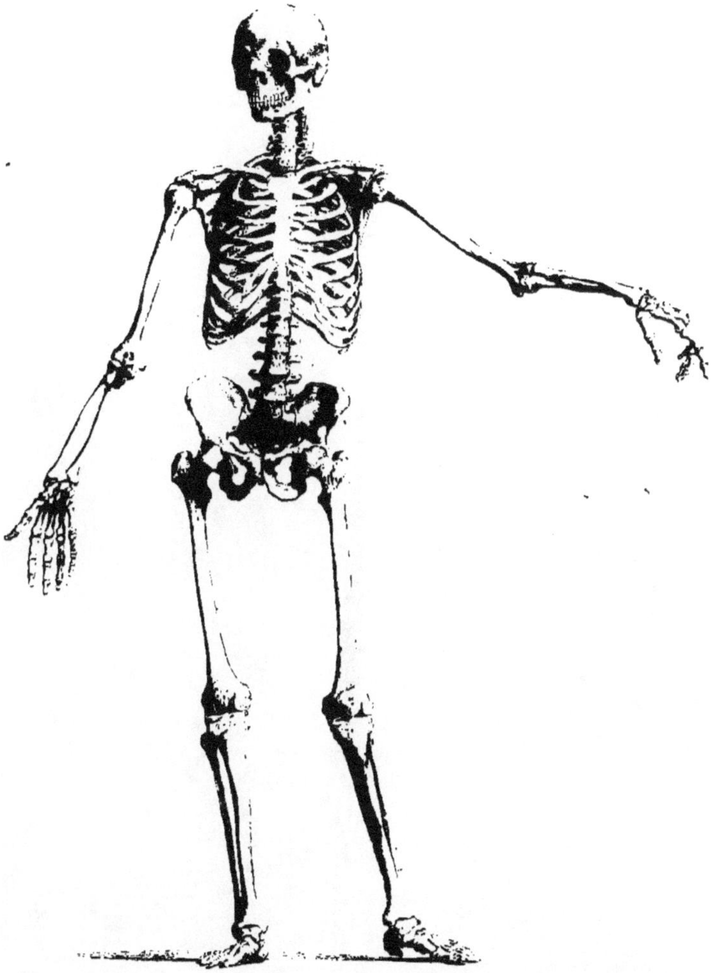

Brodum M.D. delineavit et edidit

THE EXPLICATION OF THE FIRST ANATOMICAL TABLE OF THE HUMAN SKELETON.

This firſt table contains chiefly a front view of the human ſkeleton. Some ligaments and cartilages are added, without which the ſyſtem of the bones would be interrupted.

IN THE HEAD AND SPINE.

A The frontal bone.

B B The ſuperciliary holes; the left one is entire, the right is only a notch, and ſo partly defective.

C D The coronal ſuture, C here it is a true ſuture, D here only ſquamous.

E The left parietal bone.

F The ſquamous ſuture, made by the conjunction of the parietal bone with the ſquamous part of the temporal.

G The ſquamous ſuture formed by the conjunction of the parietal with the great lateral proceſs of the multiform, ſphenoïdal, or wedge-like bone.

H The ſquamous ſuture, by the conjunction of the frontal with the ſame proceſs of the multiform bone.

I The great lateral proceſs of the multiform bone.

K The ſuture common to that proceſs and the ſquamous bone.

L The ſquamous part of the temporal bone.

M The entry into the bonny parts which compoſe the organ of hearing.

N The mammillary proceſs of the temporal bone.

O The Zygomatic proceſs of the temporal bone.

P The ſuture common to the ckeek or jugal bone, with the zygomatic proceſs of the temporal bone.

Q Q The cheek or jugal bones.

R R The ſutures common to the frontal and cheek bones near the tails of the eyebrows.

S S The ſutures which appear upon the cheeks by the conjunction of the cheek or jugal and ſuperior maxillary bones.

T T That part of the cheek bones, which aſſiſts in compoſing the ſockets of the eye.

Between T and W, the ſuture which is formed in the ſocket of the eye, by the conjunction of the cheek bone with the ſuperior maxillary.

Between T and C: T and C, the ſuture common to the cheek and frontal bones within the orbit.

Between T and Y: T and Y, the ſutures common to the cheek bones, with the great lateral proceſs of the multiform.

V V The fiſſures in the bottom of the ſockets of the eyes.

W X The part of the ſuperior maxillary bone, which compoſes the bottom of the ſocket of the eye.

Between

Between W and X, the future running along the canal, that is ftretched along the bottom of the eye-focket, which future likewife paffes over the margin of that focket, and reaches to the exit of that canal, which exit is on the cheek a little below that margin.

Between X and *d*, the future common to the fuperior maxillary bone and os planum *(d)*.

Between X and *e*, the future common to the fuperior maxillary bone and os unguis *(ef)*.

Y Y The parts of the great lateral proceffes of the multiform bone which help to compofe the fockets of the eyes.

Between Y and *c*: Y and *c*, the futures common to the great lateral proceffes of the multiform and frontal bone in the eye-fockets.

Z The hole by which the third, fourth, fixth, and firft branch of the fifth pair of nerves, &c. enter the eye-focket from the cavity of the fkull.

a The fmall procefs of the multiform bone.

b The hole by which the optic nerve, with a branch of the internal carotid artery, enters the eye-focket from the cavity of the fkull.

Between *a* and *c*, the future common to the fmall procefs of the multiform and frontal bones, within the eye-focket.

Between *a* and *d*, the future common to the fmall procefs of the multiform bone and os planum, within the eye-focket.

c c The parts of the frontal bone that help to compofe the fockets of the eyes.

Between *c* and *d*, the future common to the frontal and plain bone.

Between *c* and *ef*, the future common to the frontal and nail bone.

d The plain bone.

Between *d* and *e*, the future common to the plain and nail bone.

ef The nail bone, *f* the groove leading to the nafal canal.

Between *f* and *g*, the future common to the nail bone and nafal procefs of the fuperior maxillary.

g g The nafal proceffes of the fuperior maxillary bones.

Between *g* and *k*: *g* and *k*, the futures common to the nafal proceffes of the fuperior maxillary and nafal bones.

h The future common to the nafal procefs of the fuperior maxillary and frontal bones.

i i The futures common to the nafal bones and the frontal.

k k The nafal bones.

Between *k* and *k*, the future common to the nafal bones.

l The interior part of the nafal procefs of the fuperior maxillary bone, belonging to the nofe.

m m The inferior fpongious bones.

Between *l* and *m* of the right fide, the future formed by the conjunction of the inferior fpongious bone with the fuperior maxillary.

n o The plate of the cribriform or fieve-like bone, which helps to compofe the partition of the nofe; *o* its extremity, to which is continued the cartilaginous part of that partition.

p The vomer or plow-fhare bone.

Between *n* and *p*, a kind of future made by the connexion of the vomer with the lamina of the fieve-bone.

q The part of the fuperior maxillary bone that belongs to the inferior part of the nofe.

r The future common to the fuperior maxillary bones.

s s The fuperior maxillary bones where they form the cheeks.

t t The holes or exit of the canals that run along the inferior part of the eye-fockets.

u The aliform procefs of the multiform bone.

w x y z The lower jaw, *x* the hole or exit of the nerve and veffels from the canal in the lower jaw; *y* the coronoid or fharp procefs; *z* the neck above which is the little head articulated with the temporal bone.

α The cartilaginous lamella in the joint of the lower jaw with the temporal bone.

β γ δ ε ζ η θ ι: *β γ δ ε ζ η θ ι*, The left teeth in each jaw, *β β* the firft incifors, *γ γ* the fecond incifors. *δ δ* the canini or dog teeth, *ε ε* the firft molares or grinders, *ζ ζ* the fecond, *η η* the third, *θ θ* the fourth, *ι ι* the fifth. The right teeth anfwering to thefe are eafily underftood.

IN THE SPINE.

κ The body of the atlas or firſt vertebra, where it reſts upon the epiſtropheus and ſupports the head.

λ The body of the epiſtropheus or ſecond vertebra, where it ſupports the atlas.

μ The inferior oblique proceſs of the fifth vertebra of the neck.

ν ξ ο π The fourth vertebra of the neck, ν the ſuperior oblique proceſs, ξ the inferior oblique proceſs, ο the tranſverſe proceſs, π the body.

ρ The hole between the third and fourth.

σ σ, &c. The ligaments between the bodies of the vertebræ that tie them one to another.

τ υ υ φ φ χ The third vertebra of the neck, τ the body, υ υ the tranſverſe proceſſes, φ φ the ſuperior oblique proceſſes, χ the inferior oblique.

ψ ψ ω ω Γ The ſecond vertebra of the neck, ψ ψ the ſuperior oblique proceſſes, ω ω the tranſverſe proceſſes, Γ the body.

Δ Δ Θ Θ Λ Λ Ξ The firſt vertebra of the neck, Δ Δ the ſuperior oblique proceſſes, Θ Θ the tranſverſe, Λ Λ the inferior oblique, Ξ the body.

Π Π Σ Φ Φ Ψ The twelfth vertebra of the back, Π Π the ſuperior oblique proceſſes, Σ the tranſverſe, Φ Φ the inferior oblique, Ψ the body.

IN THE SPINE, THORAX, CLAVICLES, SCAPULA, SHOULDERS.

Ω a a b b The eleventh vertebra of the back, Ω the body, a a the ſuperior oblique proceſſes, b b the tranſverſe.

c The tranſverſe proceſs of the ſixth of the back.

d d e e The third vertebra of the back, d d the body, e e the tranſverſe proceſſes.

f g g The ſecond vertebra of the back, f the body, g g the tranſverſe proceſſes.

h The body of the firſt vertebra of the back.

i k k The fifth vertebra of the loins, i the body, k k the tranſverſe proceſſes.

l m m n The fourth of the loins, l the body, m m the tranſverſe proceſſes, n the ſuperior oblique.

o p p The third of the loins, o the body, p p the tranſverſe proceſſes.

q q r r s The ſecond of the loins, q q the ſuperior oblique proceſſes, r r the tranſverſe, s the body.

t t u u v v w The firſt of the loins, t t the ſuperior oblique proceſſes, u u the tranſverſe, v v the inferior oblique, w the body.

x x y y z z z : z z z : A A A A The os ſacrum, x x the ſuperior oblique proceſſes of its firſt vertebra, y y the ſides of the os ſacrum, z z z : z z z the three firſt holes on the fore ſide right and left, A A A A the four ſuperior bodies, between which are the bony lines that were formerly ligaments.

B The fourth little bone of the coccyx.

C D E F the ſternum or breaſt bone, C the upper portion, D the middle one, E the inferior, or that connected with the ſword-like cartilage, ſo called, F the ſword-like cartilage.

G H The ligaments by which the bones of the ſternum are bound together, G by which the middle with the inferior bone, H by which the middle with the ſuperior.

I K L M : I K L M The firſt pair of ribs, K the little head by which it is articulated with the tranſverſe proceſs of the twelfth vertebra of the back, L the beginning by which it is articulated with the body of the ſame vertebra, M the cartilaginous extremity by which it is continued with the ſternum.

N N O P : N N O P The ſecond pair of ribs, O the beginning by which it is joined with the bodies of the 11th and 12th vertebræ of the back, P the griſtly extremity.

Q Q Q R : Q Q Q R The third pair of ribs, R the griſtly extremity.

S S S T : S S S T The fourth pair of ribs, T the griſtly extremity.

V V V V W X : V V V V W The fifth pair of ribs, W the griſtly extremity, X here it becomes broad and is joined to the cartilage of the ſeventh rib to which it reaches.

Y Y Y Y Z Γ : Y Y Y Y Z Γ The ſixth pair of ribs, Z the griſtly extremity, Γ becoming broad at this part and connected to the cartilage of the ſeventh rib to which it reaches.

Δ Δ Δ Θ Λ : Δ Δ Δ Θ Λ The ſeventh pair of ribs, Θ the griſtly extremity, Λ here it becomes broad and is joined to the cartilage of the eighth rib.

Ξ Ξ Ξ Ξ Π Σ : Ξ Ξ Ξ Π The eighth pair of ribs, Π the griſtly extremity, Σ at this part becoming broad in ſome ſubjects, and reaching to the cartilage of the ſeventh rib and united to it.

Φ Φ Φ Φ Ψ : Φ Φ Φ Ψ The ninth pair of ribs, Ψ the griſtly extremity.

Ω Ω Ω Ω Ω α : Ω Ω Ω Ω Ω α The tenth pair of ribs, α the griſtly extremity.

β β β β γ : β β β β γ The eleventh pair of ribs, γ the griſtly extremity.

δ ι : δ ι The twelfth pair of ribs, ι the griſtly extremity.

ζ η ι : ζ η ι The clavicles or collar-bones, η the head that reſts upon the ſternum, ι the head that reaches to the ſuperior proceſs of the ſcapula.

θ θ The cartilages in the articulations of the clavicles with the ſternum.

κ κ The cartilages in the articulations of the clavicles with the ſuperior proceſſes of the ſcapulæ or ſhoulder blades.

λ λ λ λ λ λ λ λ λ μ ν ξ ο π : λ λ λ λ λ λ λ λ λ λ μ ν ξ ο π The ſcapulæ or ſhoulder blades, μ the ſpine, ν the coracoïd or crow bill proceſs, ξ the acromion or ſuperior proceſs, ο the neck, π the cartilaginous cruſt, by which the neck is augmented.

IN THE ARMS, FORE ARMS AND HANDS.

ρ σ τ υ φ χ ψ ω a b : ρ σ τ υ φ χ ψ ω a b The arm bones, ρ the head covered with a ſmooth cartilage, σ the larger unequable ſwelling of the ſuperior head, τ the leſſer unequable ſwelling of the ſame head, between theſe ſwellings is the ſinus in which ſlides the tendon of the longer head of the biceps muſcle, φ the ſinus that receives the ſuperior head of the radius when the fore arm is fully bent, χ the ſinus that receives the proceſs of the ulna when the fore arm is fully bent, ψ a rounded head incruſted with a ſmooth cartilage by which it is articulated to the ulna, ω a tubercle incruſted with a ſmooth cartilage by which it is articulated to the radius, *a* the leſſer condyle, *b* the greater.

c d e f g : *c d f g* The two ulnæ, *d* its ſuperior head, with the unequable ſurface into which the brachialis internus muſcle is inſerted, *e f* its little head which ſupports the radius below, and there the ſurface *f* covered with a ſmooth cartilage, *g* the ſtyloïd proceſs.

h i n o p q : *h k l m* The two radii, *i* the ſuperior head, *k* the ſurface of that head covered with a ſmooth cartilage, *l* the little head, at the poſterior part of which is inſerted the tendon of the biceps muſcle, this is turned forward in the pronation of the hand, *m n o p q* the inferior heads, *n* the ſinus in which ſlide the tendons of the long abductor, and the leſſer extenſor of the thumb, *o* the ſinus for the tendon of the radialis externus longior, *p* the ſinus for the tendon of the radialis externus brevior, *q* the ſinus for the tendon of the greater extenſor of the thumb.

r s t : *r t* The oſſa navicularia, or the navicular bones of the carpus, *s* the protuberance articulated to the radius, covered with a ſmooth cartilage, *t* the protuberance articulated with the multangular bones, covered likewiſe with a ſmooth cartilage.

u w : *u* The oſſa lunata or lunated bones, *w* the tubercle covered with a ſmooth cartilage, whereby it is articulated to the radius.

x x The oſſa triquetra or triangular bones. In the right one a ſmooth cartilaginous cruſt whereby it is articulated to the lunated bone and to the ulna.

y The os ſubrotundum, or roundiſh bone.

z A A B : z A The cuneiform or wedge-like bones of the carpus, A A the part covered with a ſmooth cartilage, articulated to the trianglar and lunated bones, B the unciform or hook-like proceſs.

C D : C D The oſſa capitata, D the head covered with a ſmooth cartilage, articulated to the lunated and navicular bones.

E E The ſmaller multangular bones.

F F The greater multangular bones.

G H : G H The metacarpal bones of the thumbs, H the inferior head covered with a ſmooth cartilage, where it is joined with the firſt bone of the thumb, and with the ſeſamoïdal bones.

I K The ſeſamoïdal bones placed at the joint of the thumb with its metacarpal bone.

L M : L M The firſt bones of the thumb, M the inferior head covered with a ſmooth cartilage, where it is joined with the ſecond and laſt bone.

N The ſeſamïodal bone placed at the laſt joint of the thumb.

O O The laſt bones of the thumbs.

P Q R S T : P Q R S T The metacarpal bones of the hands, P the metacarpal bone of the index or fore finger, Q the ſame bone of the middle finger, R of the ring finger, S of the little finger, T the inferior head covered with a ſmooth cartilage by which it is articulated with the finger; and the ſame in the reſt.

V W Small ſeſamoïdal bones found in ſome ſubjects.

X Y Z Γ Δ : X Y Y Z Z Γ Δ The firſt phalanges of the fingers, X that of the index, Y of the mid-figure, Z of the ring-finger, Γ of the little-finger, Δ the inferior head covered with a ſmooth cartilage articulated with the ſecond phalanx, and ſo in the other fingers.

Θ Λ Ξ Π Σ : Θ Λ Ξ Π Σ The ſecond phalanges of the fingers, Θ that of the index, Λ of the mid-finger, Ξ of the ring-finger, Π of the little-finger, Σ the inferior head covered with a ſmooth cartilage, articulated with the third phalanx, and ſo in the other fingers.

Φ Ψ Ω α : Φ Ψ Ω α The third phalanges of the fingers, Φ that of the index, Ψ of the mid-finger, Ω of the ring-finger, α of the little finger.

IN THE HAUNCHES AND FEET.

β γ δ ε ε ζ η θ ι ι κ λ μ : β γ δ ε ε ζ η θ ι ι κ λ μ The oſſa coxarum or haunch bones, β γ δ the os ilium, γ its criſta, δ the tubercle from which the rectus cruris muſcle riſes, ε ε ζ the iſchion, ζ the ſinus through which paſſes the iliacus internus and great pſoas muſcles, η the acute proceſs of the iſchion, θ the protuberance of the iſchion, ι ι κ λ the os pubis or ſhare bone, κ the ſpine of the os pubis, from which riſes the pectineus muſcle, λ the tubercle in which is inſerted the inferior and exterior tendon of the double aponeuroſis of the external oblique muſcle of the abdomen, μ the great foramen or hole.

ν The cartilage inſerted between the oſſa pubis, and connecting them together.

ξ ο π ρ σ τ υ φ χ ψ : ξ ο π ρ σ τ υ φ χ χ ψ ψ The thigh bones, ξ the head covered with a ſmooth cartilage, ο the neck, π the great trochanter, ρ the rough eminence to which the ligament is affixed which ſecures the joint of the haunch, σ the leſſer trochanter, υ the exterior condyle, φ the interior, χ the ſinus cruſted with a ſmooth cartilage belonging to the articulation of the patella, ψ to this part reaches the ſmooth cartilaginous cruſt which covers the condyles where they are articulated with the tibiæ.

ω ω The patellæ or knee pans.

a b : *a b* The interior ſemilunar cartilages of the joints of the knees, at *b b* they at laſt become ligaments, and are inſerted in the tibiæ.

c d : *c d* The exterior ſemilunar cartilages, *d d* at laſt they become ligaments and are inſerted in the tibiæ.

e f g h i k l : *e f g h i k l* The tibiæ, *e* the upper head, *f g* the ſmooth cartilaginous cruſts covering the tops of the tibiæ at the joint of the knee, *h* the tubercle to which is affixed the ligament proceeding from the patella and joining it to the tibia, *i* the ſpine, *k l* the lower head, *l* the inner ancle.

m n o : *m n o* The fibulæ, *m* the upper head, *n* the ſpine, *o* the lower head which is the outer ancle.

p q r : *p q r* The bones called tali, *q* the ſmooth cartilaginous cruſt with which its protuberance is covered, where it is articulated with the leg, *r* a like cruſt with which its head is covered.

s t : *s t* The heel bones, *t* the part that ſupports the neck of the talus.

u u The navicular bones of the tarſi.

v v The great cuneiform bones of the tarſus.

w w The ſmall cuneiform bones of the tarſus.

x x The middle-ſized cuneiform bones of the tarſus.

y The cubiform bone.

z A B C D E : *z* A B C D E The bones of the metatarſus, *z* that of the fourth ſmall toe, or little toe, A that of the third, B of the ſecond, C of the firſt, D of the great toe, E the ſmooth cartilaginous cruſt with

with which the head is covered that is articulated with the firſt bone of the great toe, and the like in the other toes.

F G: F The ſeſamoïdal bones placed at the joints of the great toe, with their metatarſal bones.

H H The firſt bones of the great toes.

I I The laſt bones of the great toes.

K L M N: K L M N The firſt phalanges of the ſmall toes, K that of the firſt, L of the ſecond, M of the third, N of the fourth.

O P Q R: O P Q R The ſecond phalanges of the ſmall toes, O that of the firſt, P of the ſecond, Q of the third, R of the fourth.

S T V W: S T V W The third phalanges of the ſmall toes, S that of the firſt, T of the ſecond, V of the third, W of the fourth.

TAB. II

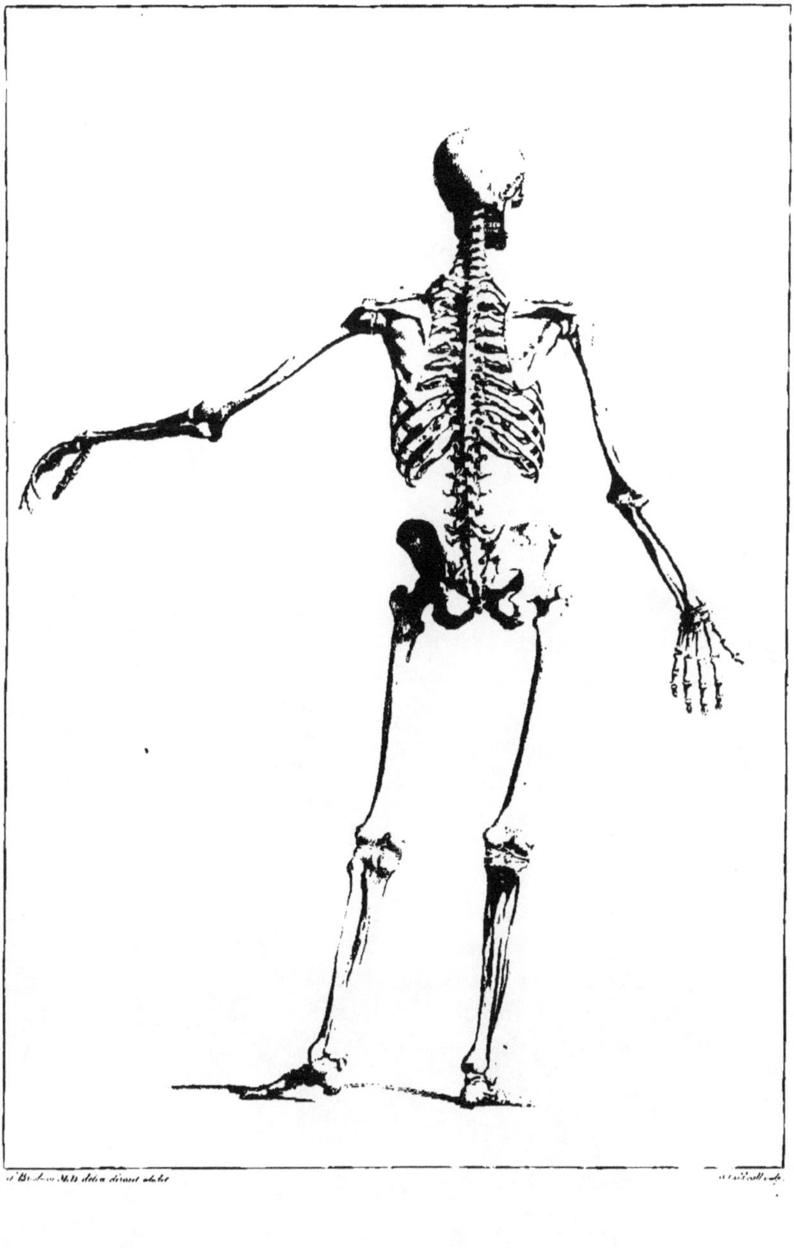

[illegible] delin. [illegible] — [illegible] sculp.

THE
EXPLICATION
OF THE
SECOND ANATOMICAL TABLE
OF
THE HUMAN SKELETON.

A back view of the ſame ſkeleton in the ſame poſition. Some ligaments and cartilages are added, in order to preſerve the connexion.

IN THE HEAD AND SPINE.

a a The parietal bones.

b b The holes in theſe bones.

c The ſagittal ſuture.

d d The lamdoïd ſuture.

e e The occipital bone.

f The ſquamous ſuture made by the conjunction of the ſquamous bone with the parietal.

g g True ſutures made by the conjunction of the mammary bones with the parietal.

h The ſquamous bone.

i i : i i The additamenta or ſupplements to the lamdoïd ſuture.

k k The holes, thro' which the branches of the internal jugular veins penetrate to the lateral ſinuſſes of the dura mater.

l l The mammillary proceſſes of the temporal bone.

m The frontal bone.

n The ſuture formed by the conjunction of the jugal or cheek bone with the frontal bone, near the extremity of the eye-brow.

o The ſuture formed by the conjunction of the zygomatic proceſs of the temporal bone with the jugal or cheek bone.

p p The jugal or cheek bone.

q The zygomatic proceſs of the temporal bone.

r The ſuperior maxillary bone.

Between *r* and the neareſt *p*, is the ſuture formed by the conjunction of the jugal with the ſuperior maxillary bone.

s s The cartilaginous lamella placed in the articulation between the lower jaw and temporal bone.

t u u u u The lower-jaw, *t* the little head, by which it is articulated with the temporal bone.

w w The parts of the ſuperior maxillary bones, that belong to the palate. In both jaws the teeth appear.

x x The ſtyliform proceſſes of the temporal bones.

y y z A B C D D E The atlas; *y y* its tranſverſe proceſſes, z the hole of the tranſverſe proceſs, A the arch which is ſometimes found: this with the ſinus by which the vertebral artery paſſes behind the body of the

H atlas,

atlas, forms the hole B, by which that artery passes into the vertebral canal, as it does at other times by the sinus in the left side C; C the sinus in the back part of the body of the atlas, where that body supports the head, thro' which sinus the artery passes in its way to the vertebral canal; D D the inferior parts of the body, where it rests upon the epistropheus; E the rough eminence instead of a spinal process, and from which arise the recti postici minores muscles of the head.

F G H H I I K K L The epistropheus, F its axis, G the interior part of the body, H H the two upper parts of the body that support the atlas, I I the transverse processes, K K the inferior oblique processes, L the spinal process.

M N Two vertebræ of the neck, M the fifth, N the fourth.

O P P Q Q R R The third vertebra of the neck, O the spine, P P the superior oblique processes, Q Q the inferior oblique processes, R R the transverse processes. From this the parts of the rest of the neck may be known.

S T Two vertebræ of the neck, S the second, T the first.

V W X Y Vertebræ of the back, V the twelfth, W the eleventh, X the tenth, Y the ninth.

Z Z α α β β γ The eighth vertebra of the back, Z Z the superior oblique processes, α α the transverse processes, β β the inferior oblique processes, γ the spinal. From this the parts of the rest of the back may be known.

δ ε ζ η θ ι Vertebræ of the back, δ the seventh, ε the sixth, ζ the fifth, η the fourth, θ the third, ι the second.

κ λ λ μ μ μ The first vertebra of the back, whose transverse processes λ λ are, as it were, reflected behind the articulation of the superior oblique processes of this vertebra, with the inferior oblique of the second; μ μ μ the body.

ν The fifth vertebra of the loins.

ξ ο ο π π ρ ρ σ σ σ The fourth vertebra of the loins, ξ the spine, ο ο the superior oblique processes, π π the transverse processes, ρ ρ the inferior oblique, σ σ σ the body. From this it will be easy to know the parts of the other lumbar vertebræ.

τ υ φ Vertebræ of the loins, τ the third, υ the second, φ the first.

χ χ, &c. The ligaments between the bodies of the vertebræ which join them to each other.

ψ ψ ω ω ω ω ω ω ω ω Γ Γ Γ Δ Δ Θ Λ The os sacrum, ψ ψ the superior oblique processes, ω ω, &c, the posterior holes, Γ Γ Γ the spines, Δ Δ the inferior oblique processes which are joined to the superior oblique of the first little bone of the coccyx, Θ the body of the fourth vertebra, of which the os sacrum is composed.

Between Θ and Λ, is the bony part, intervening between, and continued with the bodies of the fourth and and fifth, formerly a ligament, Λ the body of the fifth vertebra of the os sacrum.

Between Λ and Ξ, the ligament intervening between the os sacrum and coccyx, and binding them together.

Ξ Π Σ The first bone of the coccyx; Ξ the body, Π the transverse process: the like too on the other side. Σ the superior oblique: the like also on the other side.

Between Ξ and Φ, the ligament, by which the first and second bones of the coccyx are bound together.

Φ Ψ Ω The little bones of the coccyx, Φ the second, Ψ the third, Ω the fourth.

IN THE THORAX, CLAVICLES, SCAPULÆ, ARMS, HANDS.

a a b b The sternum or breast bone.

c c d d : c c d d The first pair of ribs, *d d* the cartilaginous part.

e e e f : e e e f The second pair of ribs, *f* the cartilaginous part.

g g h : g g g h The third pair of ribs, *h* the cartilaginous part.

i i k : i i k The fourth pair of ribs, *k* the cartilaginous part.

l l m : l l m The fifth pair of ribs, *m* the cartilaginous part.

n n n n o o : n n n n n o The sixth ribs, *o* the cartilaginous part.

p p p p p q q : p p p p q q The seventh ribs, *q* the cartilaginous part.

r r r s : r r r r s The eighth ribs, *s* the cartilaginous part.

t t t u : t t u The ninth ribs, *u* the cartilaginous part.

w x x : w x x x The tenth ribs, *x* the cartilaginous part.

y z : *y z* The eleventh ribs, *z* the cartilaginous part.

A B : A B The twelfth ribs, B the cartilaginous part.

C C C C : C C C The clavicles.

D D The cartilaginous lamellæ, intervening between the articulations of the clavicles, with the superior processes of the scapulæ.

E E F G H I : E E F G H I The scapulæ, F the spine, G the superior process, H the neck, I the cartilage that incrusts the sinus of the neck.

K L M N O P : K L M N O P The arm bones, K the head covered with a smooth cartilage, where it is articulated with the sinus of the scapula, L the larger unequable tubercle of the superior head, M the sinus, along which are conveyed an artery, vein, and nerve, N the sinus which receives the olecranon when the fore arm is extended, O the lesser condyle, P the greater.

Q R S T : Q R S T The ulnæ, R the olecranon, S the little head by which it supports the radius below, T the styloid process.

V V W X : V W X Y Z *α β* The radii. W X the superior little head, X the circumference of this little head covered with a smooth cartilage, by which it is moved on the sinus of the ulna, Y the sinus that contains the tendons of the obductor longus and extensor minor muscles of the thumb, Z the sinus that contains the tendons of the radiales externi, *α* the sinus for the tendon of the extensor major of the thumb, *β* the sinus for the tendons of the common extensor of the fingers, of the proper extensor of the little finger, and of the indicator mucle.

γ δ ε, *γ* The navicular bones of the carpus, *δ* the little head covered with a smooth cartilage, by which it is articulated to the radius, *ε* the little head covered with a smooth cartilage, by which it is articulated with the multangular bones.

ζ ζ The lunated bones; that of the right hand, where it is articulated with the radius, covered with a smooth cartilage.

η θ, *η* The triangular bones, *θ* the surface covered with a smooth cartilage, where it is articulated with the cuneiform.

ι ι The roundish bones.

κ λ : *κ λ* The cuneiform bones of the carpus, *λ* the surface covered with a smooth cartilage, by which it is articulated with the triangular.

μ ν, *μ* The ossa capitata, *ν* the head covered with a smooth cartilage, by which it is articulated to the navicular and lunated bones.

ξ ξ The lesser multangular bones.

ο ο The greater multangular bones.

π ρ, *π* The metacarpal bones of the thumbs, *ρ* the inferior head covered with a smooth cartilage, where it is articulated with the first bone of the thumb, and with the sesamoidal bones: the same in the left thumb.

σ σ The sesamoid bones placed at the articulation of the thumb with its metacarpus.

τ υ : *τ υ* The first bones of the thumbs, *υ* the smooth cartilaginous crust, which covers that part of the inferior head that is articulated to the last bone of the thumb.

φ φ The last bones of the thumb.

χ ψ ω Γ Δ : *χ ψ ω* Γ Δ The metacarpal bones, *χ* of the index, *ψ* of the middle finger, *ω* of the ring finger, Γ Δ of the little finger, Δ the cartilaginous crust covering the inferior head, whereby it is articulated with the first phalanx: the same in the rest.

Θ Λ Ξ Π Σ : Θ Λ Ξ Π The first phalanges of the fingers, Θ of the little finger, Λ of the ring, Ξ of the middle, Π Σ of the index, Σ the cartilaginous crust covering the inferior head, by which it is articulated with the second phalanx: the same in the others.

Φ Ψ Ω *a b* : Φ Ω *a* The second phalanges of the fingers, Φ of the index, Ψ of the middle finger, Ω of the ring, *a b* of the little finger, *b* the part of the inferior head covered with a smooth cartilage to articulate with the third phalanx: the like in the rest.

c d e f : *c d e f* The third phalanges of the fingers.

IN THE HAUNCHES AND INFERIOR EXTREMITIES.

g h i k l m m : *g h i i k l m* The offa coxarum or haunch bones, *g h* the os ilium, *h* the crifta, *i* the ifchion, *k* the fharp procefs of the ifchion, *l* the tuberofity of the ifchion, *m* the os pubis or fhare bone.

n o p q r s t u w x : *o p q r s t u w x* The thigh bones, *n* the head feated in the acetabulum and covered with a fmooth cartilage, *o* the neck, *p* the greater trochanter, *q* the leffer, *r* the linea afpera or rough eminence ftretched along the back part of the thigh bone, *s t* the outer condyle, of which the part *t*, belonging to the joint of the knee, is covered with a fmooth cartilage, *u w* the inner condyle, *w* where it belongs to the joint of the knee covered with a fmooth cartilage, *x* the finus between the condyles.

y y The exterior femilunar cartilages, which are inferted in the joints of the knees, and becoming ligaments *z z* are at laft inferted in the interior condyles.

A A The interior femilunar cartilages inferted in the joints of the knees, B B their extremities becoming ligaments are fixed in the tibiæ.

C D E F G : C D E F G The tibiæ, D E the parts of the fuperior head belonging to the joint of the knee covered with a fmooth cartilage, F the internal ancle, G the finus thro' which paffes the tendon of the tibialis pofticus, and long flexor of the toes.

H I K L : H I K L The fibulæ, I the upper head, by which it is joined to the tibia, K the external ancle, L the finus thro' which pafs the tendons of the two peronei, the long and fhort.

M N O P : M N O P The tali. N O the fmooth cartilage, with which its tuberofity being covered is articulated with the tibia N, and with the fibula O ; P the head.

Q R : Q R The heel bones, R the eminence, about which is ftretched the tendon of the peroneus longus

S S The navicular bones of the tarfus.

T T The leffer cuneiform bones of the tarfus.

V V The middle cuneiform bones of the tarfus.

W W The cubiform bones.

X Y Z *α* : X X Y Z *α* The metatarfal bones, X that of the firft of the fmall toes, Y of the fecond, Z of the third, *α* of the fourth.

β γ δ ε : *β γ δ* The firft phalanges of the fmall toes, *β* of the fourth, *γ* of the third, *δ* of the fecond, *ε* of the firft.

ζ η : *ζ* The fecond phalanges of the fmall toes, *ζ* of the fourth, *η* of the third.

θ ι κ : *θ* The third phalanges of the fmall toes, *θ* of the fourth, *ι* of the third, *κ* of the fecond.

λ The firft bone of the great toe.

μ The metatarfal bone of the great toe.

ν The greater cuneiform bone of the tarfus.

ξ ο The fefamoid bones placed at the articulation of the great toe with its metatarfus, *ξ* the internal one, *ο* the external.

f

TAB. III

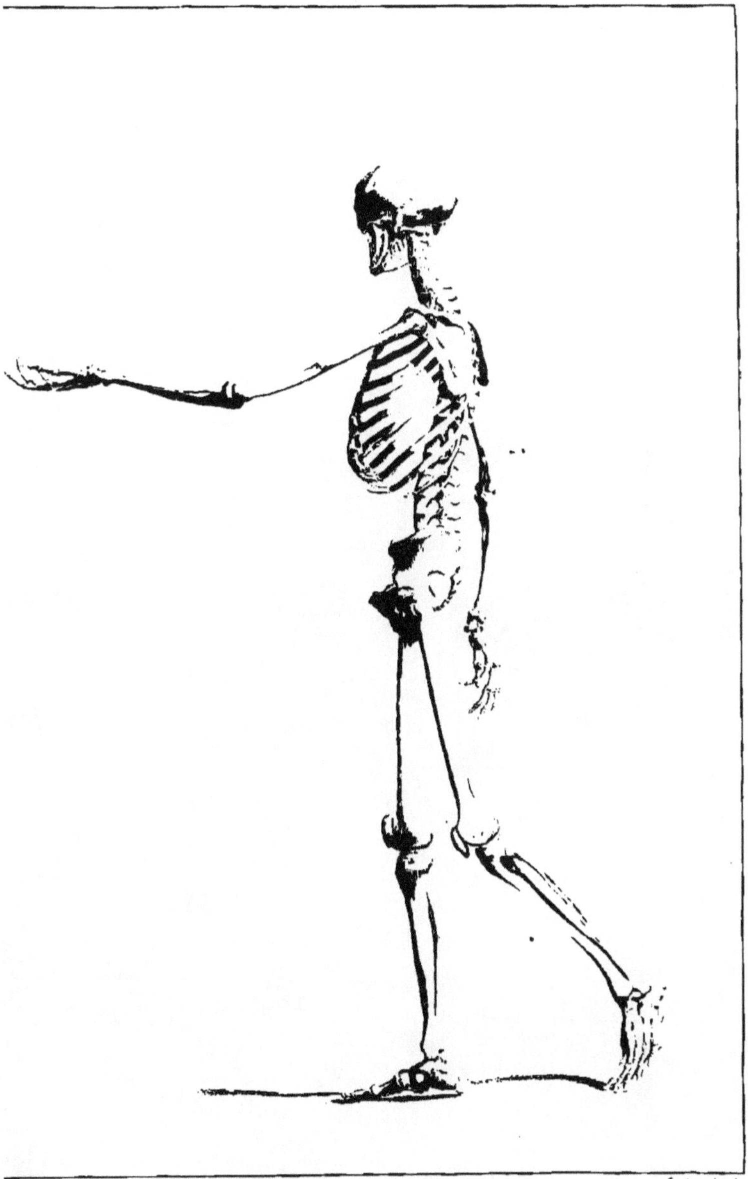

THE

EXPLICATION

OF THE

THIRD ANATOMICAL TABLE

OF

THE HUMAN SKELETON.

This likewiſe repreſents the ſame ſkeleton in a ſide view, but in a different poſition. To this are alſo added ſome ligaments and cartilages neceſſary to preſerve the connexion.

IN THE HEAD AND SPINE.

A A The parietal bones.

B The ſagittal ſuture.

C C The holes in the parietal bones.

D D The lamdoïd ſuture.

E The occipital bone.

F G : G The mammillary proceſſes of the temporal bones, F the eminence from which the biventer muſcle of the lower jaw riſes.

H The holes, one in the mammillary bone near the appendage of the lambdoïd ſuture, the other in that appendage itſelf; thro' which hole a vein paſſes to the lateral ſinus of the dura mater.

I The appendage of the lamdoïd ſuture.

K A true ſuture made by the conjunction of the mammillary bone with the parietal.

L The mammillary bone.

M The bony entrance to the ear.

N The zygomatic proceſs of the temporal bone.

O The ſquamous bone.

P The ſquamous ſuture; made by the conjunction of the ſquamous bone with the parietal.

Q R S The coronal ſuture, Q in this part it is a true ſuture, R S here it is a ſquamous one, where the frontal bone rides at R upon the parietal, and at S upon the multiform bone.

T The frontal bone.

V The ſquamous ſuture, made by the conjunction of the multiform bone and the parietal.

W The ſuture, formed by the conjunction of the great lateral proceſs of the multiform and the ſquamous bone.

X The great lateral proceſs of the multiform.

Y The ſuture, common to the frontal and jugal or cheek bone, near the extremity of the eye-brow.

Z That part of the jugal bone that lies in the hollow of the temples.

Below Z is the ſuture common to the jugal and ſuperior maxillary bone in the hollow of the temple.

Between Z and X is the ſuture, common to the jugal bone and the great lateral proceſs of the multiform.

a The ſuperior maxillary bone.

Between *a* and X the fiſſure left between the ſuperior maxillary bone, the jugal and the multiform.

I

b The exterior part of the jugal bone.

c The ſuture, common to the jugal bone and the zygomatic proceſs of the temporal bone.

d The ſuperior maxillary bone.

e f g g The lower jaw, *e* the coronoïd proceſs, *f* the condyle by which it is articulated with the temporal bone.

Directly above *f* is the cartilaginous lamella, interpoſed between theſe articulated parts of the lower jaw.

h i The concave part of the left pterygoïd proceſs of the multiform bone, *i* the little hook which ſuſtains and holds the tendon of the circumflex muſcle of the palate.

k l The ſuperior maxillary bone, *k* the part that belongs to the gums, *l* the part that belongs to the palate.

m m m The teeth in both jaws.

n o o p p q The atlas, *n* the left part of the body where it receives the coronoïd proceſs of the occipital bone, and ſuſtains the head by a moveable joint, *o o* the two inferior parts of the body by which it reſts upon the epiſtropheus by moveable joints, *p p* the tranſverſe proceſſes, *q* the inequality in place of a ſpine, from which ariſe the recti poſtici minores of the head.

r r s t u The epiſtropheus, *r r* two parts of the body, by which it ſupports the atlas by moveable joints, *s* the tranſverſe proceſs, in which is the hole for the vertebral artery and vein, *t* the inferior oblique proceſs, *u* the ſpine forked at the extremity.

v w x y z The fifth vertebra of the neck, *v* the body, *w* the tranſverſe proceſs, *x* the ſuperior oblique proceſs, *y* the inferior oblique, *z* the ſpine.

α α, &c. The ligaments between the bodies of the vertebræ which bind them together.

β γ δ ε Vertebræ of the neck, β the fourth, γ the third, δ the ſecond, ε the firſt: their ſeveral parts may be known by theſe of the fifth vertebra.

ζ η θ The twelfth vertebra of the back, ζ the body, η the tranſverſe proceſs, θ the ſpine.

ι κ λ The eleventh vertebra of the back, ι the tranſverſe proceſs, λ the ſpine.

μ The tranſverſe proceſs of the tenth vertebra of the back.

ν ν, &c. The paſſages between the vertebræ for the ſpinal nerves, &c.

ξ ο π ρ σ The ſpines of the vertebræ of the back, ξ of the tenth, ο of the ninth, π of the eighth, ρ of the ſeventh, σ of the ſixth.

τ υ The fifth vertebra of the back, τ the ſpine, υ the body.

φ χ ψ The fourth vertebra of the back, φ the body, ψ the ſpine.

ω Γ Δ The third vertebra of the back, ω the body, Γ the inferior oblique proceſs, Δ the ſpine.

Θ Θ Λ Ξ The ſecond vertebra of the back, Θ Θ the body, Λ the ſuperior oblique proceſs, Ξ the ſpine.

Π Σ Φ The firſt vertebra of the back, Π the body, Φ the ſpine.

Ψ Ψ Ω The fifth vertebra of the loins, Ψ Ψ the body, Ω the ſpine.

IN THE SPINE.

A A Æ B C C D The fourth vertebra of the loins, A A the body, Æ the ſuperior oblique proceſs, B the tranſverſe, C C the inferior oblique, D the ſpine.

E F G H I The third vertebra of the loins, E the body, F the tranſverſe proceſs, G the ſuperior oblique, H the ſpine, I the inferior oblique.

K The ſecond vertebra of the loins, its parts are known by the former.

L M The firſt vertebra of the loins: L the ſuperior oblique proceſs, M the ſpine.

N O P The os ſacrum, N the unequable lateral part below the os ilium, O the third ſpine, P the inferior oblique proceſs, articulated with the ſuperior oblique of the firſt bone of the coccyx.

Q R The firſt bone of the coccyx, Q the ſuperior oblique proceſs, R the body.

S T The little bones of the coccyx, S the ſecond, T the third.

IN THE THORAX, SCAPULÆ, CLAVICLES.

V W X The firſt rib on the left ſide, V its beginning where it is articulated to the body of the twelfth vertebra of the back, W the little head articulated with the tranſverſe proceſs of the ſame vertebra.

Y Z *a a b* The ſecond rib on the left ſide, Y its beginning where it is articulated in the ſinus common to the bodies of the eleventh and twelfth vertebræ of the back, Z the little head by which it is articulated with the tranſverſe proceſs of the eleventh, *b* its cartilaginous extremity.

c The ſecond rib on the right ſide.

d d e The third rib on the left ſide, *e* its cartilaginous extremity.

f f The third rib on the right ſide.

g h The fourth rib on the left ſide, *h* its cartilaginous extremity.

i i k The fourth rib on the right ſide, *k* its cartilaginous extremity.

l m The fifth rib on the left ſide, *m* its cartilaginous extremity.

n n o o The fifth rib on the right ſide, *o o* its cartilaginous extremity.

p p q The ſixth rib on the left ſide, *q* its cartilaginous extremity.

r r s s The ſixth rib on the right ſide, *s s* its cartilaginous extremity.

t t u The ſeventh rib on the left ſide, *u* its cartilaginous extremity.

v v w w The ſeventh rib on the right ſide, *w w* its cartilaginous extremity.

x y The eighth rib on the left ſide, *y* its cartilaginous extremity.

z z z, 1, 1 The eighth rib on the right ſide, 1, 1 its cartilaginous extremity.

2, 3 The ninth rib on the left ſide, 3 its cartilaginous extremity.

4, 4, 4, 5, 5 The ninth rib on the right ſide, 5, 5 its cartilaginous extremity.

6, 6, 7 The tenth rib on the left ſide, 7 its cartilaginous extremity.

8, 8, 9 The tenth rib on the right ſide, 9 its cartilaginous extremity.

10, 10, 11 The eleventh rib on the left ſide, 11 its cartilaginous extremity.

12, 12, 13 The eleventh rib on the right ſide, 13 its cartilaginous extremity.

14, 15, 15, 16 The twelfth rib on the left ſide, 14 its beginning whereby it is articulated with the body of the firſt vertebra of the back, 16 its cartilaginous extremity.

17 The inner ſide of the right ſcapula.

18, 19, 20, 21, 22 The left ſcapula, 19 the neck, 20 the cartilaginous part by which the neck is augmented and the ſinus is covered that is articulated with the head of the arm bone, 21 the ſpine, 22 the ſuperior proceſs.

23 The left clavicle.

24, 24, 24 The breaſt bone.

IN THE ARMS AND HANDS.

A B C D E F: A F G H The arm bones, A in the left, the eminence where the deltoïd muſcle ends. B C D the ſuperior head, B the leſſer rough tubercle of the ſuperior head, C the greater rough tubercle of the ſame head; between B and C, the ſinus in which is contained the tendon of the longer head of the biceps muſcle, D the ſmooth cartilaginous cruſt, with which that part of the head is covered, that is articulated with the ſinuoſity of the ſcapula, E the leſſer condyle, F the head covered with a ſmooth cartilage, to which the radius is articulated; G the circumference covered with a ſmooth cartilage, with which the ulna is articulated; H the greater condyle.

I K L: I K L M The ulnæ, I the olecranon, L the little head covered all round with a ſmooth cartilage, which is articulated with the radius; M the ſtyloïd proceſs.

N O P Q Q: N O P R S The radii, O the ſuperior head, P the tubercle at the poſterior part of which is inſerted the tendon of the biceps muſcle: this tubercle is turned forward in the pronation of the hand. Q Q: R S the inferior heads, R the ſinus, thro' which paſs the tendons of the leſſer extenſor and long abductor

abductor of the thumb, S the ſinus again divided into two, through which paſs the tendons of the radiales externi muſcles.

T V : T The navicular bones of the carpi, V the head covered with a ſmooth cartilage, by which it is articulated to the multanguli.

W W The lunated bones.

X Y The os triquetrum, X the part covered with a ſmooth cartilage, where it is articulated with the ulna; a ligament intervening, which extends from the bottom of the little head of the ulna, to the bottom of the radius, where that bone is joined to the ulna.

Z Z The roundiſh bones.

a a The greater multangular bones.

b b The leſſer multangular bones.

c c c The oſſa capitata.

d d e f The cuneiform bones of the wriſts, *e* the part covered with a ſmooth cartilage, where it is joined to the triangular bone, *f* the unciform proceſs.

g g h The metacarpal bones of the thumb, *h* the cartilaginous cruſt that covers the inferior head where it is articulated with the firſt phalanx, and joined to the ſeſamoïdal bones: the ſame in the right thumb.

i i The ſeſamoïdal bones, placed at the joint of the thumb with its metacarpus.

k k l The firſt bones of the thumb, *l* the cartilaginous cruſt, covering the inferior head where it is articulated to the laſt bone of the thumb.

m m The laſt bones of the thumbs.

n n p q r : *n o p r* The metacarpal bones of the hands; *n* of the index, *p* of the middle finger, *q* of the ring finger, *r* of the little one, *o* the ſmooth cartilaginous cruſt, covering the inferior part of the metacarpal bone of the index, where it is articulated with the firſt phalanx: and the ſame of the reſt in both hands.

s t u v : *s t u v w* The firſt phalanges of the fingers, *s* of the little finger, *t* of the ring finger, *u* of the mid-finger, *v* of index, *w* the ſmooth cartilaginous cruſt, covering the inferior head, where it is articulated to the ſecond phalanx: the like in the reſt.

x y z Γ; *x y z* Γ Δ The ſecond phalanges of the fingers; *x* the index, *y* of the middle finger, *z* of the ring finger, Γ of the little one, Δ the inferior head covered with a ſmooth cartilage, where it is articulated with the third phalanx: the ſame in the reſt.

Θ Λ Ξ Π : Θ Λ Ξ The third phalanges of the fingers, Θ of the index, Λ of the middle finger, Ξ of the ring finger, Π of the little one.

IN THE HAUNCHES AND LOWER EXTREMITIES.

Σ Φ Ψ Ω *a b* The left os coxæ or haunch bone, Σ Φ Ψ the os ilium, Φ te criſta, Ψ the tubercle, from which riſes the rectus cruris muſcle, Ω *a* the iſchion, *a* the acute proceſs, *b* the os pubis.

c d e f f The right os coxæ or haunch bone, *c* the criſta of the ilium, *d* the tubercle, from which riſes the rectus cruris, *e* the acute proceſs of the iſchion, *f f* the os pubis.

g h i k l m The left thigh bone, *g* the head, covered with a ſmooth cartilage, which is articulated with the acetabulum, *h* the neck, *i* the greater trochanter, *l* the exterior condyle, *m* thus far extends the ſmooth cartilaginous cruſt that covers the part of the condyle belonging to the joint of the knee.

n n o p p The right thigh bone, *o* the inner condyle, *p p* thus far extends the ſmooth cartilaginous cruſt that covers that part of the condyle which is articulated with the tibia and patella.

q r : *q r* The patellæ, *r* on this part, which belongs to the joint of the knee, covered with a ſmooth cartilaginous cruſt.

s s The exterior ſemilunar cartilages inſerted between the joints of the knees.

t The interior ſemilunar cartilage inſerted between the ſame joint.

u v v w x y z : *u v w x y z* The tibiæ, *u* the ſuperior head, *v* here where it belongs to the joint of the knee covered with a ſmooth cartilage, *w* the eminence where the ligament proceeding from the patella is inſerted, binding that bone to the tibia.

y z The inferior head, *z* the internal ancle.

A B C : A B C The fibulæ, B the ſuperior head, C the external ancle.

D E F G : D E G The tali, E here at the joint with the leg it is covered with a ſmooth cartilage, F the ſinus through which paſſes the tendon of the long flexor of the great toe, G the cartilaginous cruſt with which the head of the talus is covered.

H : H I K The heel bones, I the knob by which it begins, at the lower and poſterior part of which are inſerted the tendo achillis and that of the plantaris. It is bent backwards and upwards when we bend the joint of the leg with the extremity of the foot forwards, K the riſing part that ſupports the head of the talus.

L L The cubiform bones.

M M The navicular bones of the tarſus.

N The middle-ſiz'd cuneiform bone of the tarſus.

O O The leſſer cuneiform bones of the tarſus.

P P The larger cuneiform bones of the tarſus.

Q R S T V : Q S T V W The metatarſal bones, Q of the great toe, R of the firſt of the ſmall toes, S of the ſecond, T of the third, V of the fourth, W the head of the metatarſal bone of the great toe, covered with a ſmooth cartilage where it is joined with the firſt bone of that toe, and with the ſeſamoïdal bones. The ſame in the others.

X The ſeſamoïdal bones, placed at the joint of the great toe with its metatarſal bone.

Y Z α : Y Z α β γ Δ The firſt phalanges of the great and ſmall toes, Y of the great toe, Z of the firſt of the ſmall toes, α of the ſecond, β of the third, γ of the fourth, Δ the head covered with a ſmooth cartilage where it belongs to the articulation with the next bone. The ſame in the other toes.

ε ε ζ η θ The ſecond phalanges of the ſmall toes, ε of the firſt, ζ of the ſecond, η of the third, θ of the fourth.

ι ι The laſt bones of the great toes.

κ κ : κ λ μ ν The third phalanges of the ſmall toes, κ of the firſt, λ of the ſecond, μ of the third, ν of the fourth.

K

TAB. II.

THE

EXPLICATION

OF THE

FIRST ANATOMICAL TABLE

OF

THE HUMAN MUSCLES.

In this table I exhibit the external mufcles, as they appear over the whole body in this pofition, after the common integuments and tendinous vaginæ are removed, together with fome ligaments belonging to them; alfo certain portions of the fkeleton, and of other parts, as the nofe, ear, and private parts, which are not covered with mufcles.

IN THE HEAD, NECK, AND TRUNK.

a a a b b c d e f g h : d e f g h i k l The epicranius mufcle, *a a a* the middle aponeurofis between the occipital and frontal mufcles, *b b, &c.* the frontal mufcles, *b b* the points by which they begin, *c* their conjunction along the middle of the forehead, *d-e d-e* here the frontal mufcles end at the orbicular mufcles of the eye-lids, *e-f e-f* here they are bent along the eye-brows to the greater angles of the eyes in the manner of the orbiculares, *g g* the points which bend into the greater angles of the eyes, *h h* the portions that accede to the levators of the upper lip and alæ of the nofe, *i* the part that runs along the glabella and nofe, *k l* its conjunction with the compreffors of the nofe, with which it is interwoven at *k*, and is continued with them at *l*.

m m n o o p q r : m o o p The orbicular mufcles of the eye-lids, *m m* the part that incircles the circumference of the orbit, *n* the part that comes from the corrugator of the eye-brow, *o o* the part that covers the eye-lids, *p* the implication of the fibres that come from the eye-lids and meet near the leffer angle, *q r* the origin from the ligament by which the meeting of the eye-lids is joined to the nofe in the larger angle of the eye.

s The ligament by which the meeting of the eye-lids in the greater angle is joined to the nofe, and to that part of it that is formed by the fuperior maxillary bone,

t u The compreffor of the nofe, *t* its flefhy portion, *u* the aponeurofis by which the right and left are joined along the ridge of the nofe.

w x y : y The levators of the upper lip and wings of the nofe, *x* the part that proceeds to the ala along the fide of the nofe, *y* the extremity which becoming thin is loft on the upper lip.

z : A z The levators of the upper lip, A its thin'd extremity by which it vanifhes along the upper lip.

B B The portions proceeding from the orbicular of the eye-lids to the upper lip.

C C The leffer zygomatic mufcles, which become thin and vanifh along the upper lip.

D D : D The levators of the angles of the mouth, D D it is in part continued with the depreffor of the angle, and partly bends itfelf round the angle of the mouth to the under lip, and there makes the exterior part of the orbicularis of the mouth.

E F G : E The greater zygomatic mufcles, F their origin from the jugal or cheek bone, G their extremity continued with the depreffor of the angle of the mouth.

H H The nasal muscles of the upper lip. Their origin from the nose appears and the manner they join themselves to the orbicular of the mouth.

I The part of the orbicular of the mouth that is upon the upper lip, where it goes round the angle of the mouth, it receives a portion from the levator of the angle going round along with it.

K K The part of the orbicular of the mouth that is in the red margin of the lips.

L L Subtile fasciculi, that proceed partly from the greater zygomatics extending hither; partly from the depressors of the angles of the mouth as it were straying. They cross or decussate the fasciculi of the depressors of the lower lip that lie under them.

M: M N The depressors of the lower lip, N here they cross each other.

O P The levators of the chin, P *fasciculi* which they mix with the fat of the chin.

Q R R S: Q The depressors of the angles of the mouth, R R their origin from the lower jaw, S their continuation with the greater zygomatic.

T The buccinator.

V W X Y: V The masseter muscles, V the fore and exterior part, W the origin of that part from the jugal bone, X the posterior part where it is not covered by the other, Y the origin of this part from the jugal bone, and from the zygomatic process of the temporal bone.

Z The anterior muscle of the external ear.

Γ Δ The raiser up of the ear, Γ its tendinous origin where it rises from the epicranius, Δ its fleshy part.

Θ The greater muscle of the helix.

Λ The tragicus.

Ξ The lesser muscle of the helix.

Π The antitragicus.

Σ The biventer muscle of the lower jaw.

Φ The sternomastoïdeus and cleidomastoïdeus united together.

Ψ Ψ The cucullares muscles.

Ω α α α β β β γ γ δ ε ε ζ η θ: Ω α α α ζ η θ The latissimi colli or platysmo-myoïdes muscles, α α α its origin, consisting of slender and chiefly of scattered fasciculi, β β β fasciculi, that sometimes accede from the side of the neck, γ γ scattered fasciculi vanishing on the cheek by which it ends, δ a fasciculus streached along the fore part of the depressor of the angle of the mouth towards the angle of that side, ε ε the lower jaw, appearing under this thin muscle, and in the same manner ζ the sternomastoïdeus, η the cleidomastoïdeus, and θ the claviculæ appear.

ι ι The sternohyoïdei.

κ The aspera arteria, or wind-pipe.

λ μ: μ The sternomastoidei, μ the tendinous origin rising from the sternum.

ν ν The sternothyroïdei.

ξ ο ο π ρ ς ς: ξ ο ο π ρ ς ς The pectoral muscles, ο-ο the origin from the sternum, π from the cartilage of the sixth rib, ρ from that of the seventh rib by a slender thin and for some time tendinous origin, ς its cohesion with the aponeurosis of the external oblique of the abdomen, ς a portion acceding from the aponeurosis of the external oblique; here tendinous and thin, in others fleshy and thicker, and in others otherways varying.

σ σ The teres major.

τ υ φ: τ υ φ The latissimi dorsi, υ φ the heads rising, υ from the tenth rib, φ from the ninth.

IN THE TRUNK.

χ ψ ω *a b c c c c*: ψ ω *a b c c c c* The serrati magni, χ the head rising from the fifth rib, ψ from the sixth, ω from the seventh, *a* from the eighth, *b* from the ninth, *c c c c* the place of the origin of the heads from the ribs.

d e f g h i k k k k l l l m m n o o o p p p p q r r r r s t u v v w w x: *d e f g h k k k k l l l m m n o o o p p p p q r r r r s t u v v w w x* The external oblique muscles of the abdomen, *d* the fleshy part, *e* the head rising from the sixth rib, *f* from the seventh, *g* from the eight, *h* from the ninth, *i* from the tenth, *k* the place of the

the origin of the heads from the ribs, *l l l m m n o o o p p p p q r r r r ſ t t u v v w w x* the aponeuroſis, *m m* here the fleſhy part of the internal oblique appears under it, *n* here under the ſame, and likewiſe under the aponeuroſis of the obliquus internus appears the fleſhy part of the tranſverſe muſcle, *o o o* here in like manner appear the recti muſcles, *p p p p* here through the aponeuroſis appear the tendinous lines of the recti, *q* here under the ſame appears the pyramidalis, *r r r r* the linea alba in which the aponeuroſes of the external oblique muſcles croſs each other, are continued into each other, and mix with the parts behind them, *ſ* the aponeuroſis inſerted into the breaſt bone, *s* this part may be ſaid to belong either to the aponeuroſis of the external oblique, or of the pectoral muſcle, and therefore either to be inſerted in the cartilage of the ſeventh rib, or to ariſe from it, *t* the hole in the linea alba through which in the embryo paſſed the umbilical arteries and vein, and the urachus, *u* the bottom of the tendinous margin extending from the criſta of the ilium, to the pubes, *v v w w* two parts into which the aponeuroſis is divided, thence diſtinct all the way to the pubes with the appearance of tendons, whereby the fiſſure is formed through which paſſes the ſpermatic cord with the cremaſter muſcle, *x* a thinner part reaching from the one of theſe tendons to the other, and connecting them together; under which part runs the ſpermatic cord and appears faintly through it, and below it, near the pubes, the cord eſcapes through the ring of this oblique muſcle, which is ſmall, and is formed between the part *x*, the tendons *v v w w*, and the os pubis; likeways the fibres of the aponeuroſis that run in the manner of fleſhy ones croſs other fine ſcattered tendinous fibres, which appears ſufficiently in the figure: and by theſe running from the one tendon through the other is formed the part *x*.

y y The naked cords of the ſpermatic veſſels.

z z The cremaſters.

IN THE THIGHS, LEGS, AND FEET.

A A The great glutæi.

B B The graciles.

C C The great adductors of the thighs.

D D The long adductors of the thighs.

E E The pectinæi.

F F The great pſoæ.

G G The internal iliacs.

H I : H I The ſartorii, I the beginning outwardly tendinous, riſing from the criſta of the ilium.

K L : K The middle glutæi, L the origin from the criſta of the ilium.

M N O : M N O The tenſors of the vaginæ of the thighs, N their origin from the criſta of the ilium, O the the extremity, from which is cut away the tendinous portion which it joins to the vaginæ of the thighs.

P Q R S : P Q R S The vaſti externi, Q the tendinous part, R the tendinous extremity, S inſerted in the patella.

T V W X : T V W X The recti muſcles, V the tendon inſerted in the patella, W the place where it is inſerted there, X the aponeuroſis which runs from the tendon of the rectus along the fore part of the patella, and afterwards joins itſelf to the fore part of the ligament which extends from the patella to the tibia.

Y Z Γ : Y Z Γ The vaſti interni, Z the extremity of its tendon, Γ inſerted in the patella.

Δ Θ Λ : Δ Θ Λ The ligaments extending from the patella to the tibiæ, Θ the place where it riſes from the patella, Λ under this whole part it is inſerted in the tibia.

Ξ Π Σ : Ξ Π Σ The bicipites of the legs, Π Σ the extreme tendon, Π its principal part inſerted in the head of the fibula; Σ the part that extends to the tibia.

Φ Ψ Ω : Φ Ψ Ω The ſartorii, Ψ the tendon, Ω inſerted in the tibia.

α α The ſemitendinoſi.

β γ δ : β γ The gemelli, γ the tendinous part, δ the tendon.

ε ζ ζ η : ε ζ ζ η The ſolei, ζ ζ their origin from the tibia, η their tendinous ſurface.

θ ι ι κ : θ ι ι κ The long flexors of the toes, ι ι their origin from the tibia, κ the beginning of the tendon.

λ λ The tendons of the tibiales poſtici.

μ μ The tendons of the plantares.

ν ν : ν The tendons of achilles.

ξ ξ The ſolei muſcles.

ο π ρ : ο π ρ The peronei longi, π their origin from the head of the fibula, ρ the tendon riſing from the exterior part of the fleſh.

σ σ The peronei breves.

τ υ : τ υ The long extenſors of the toes united with the peronei tertii, υ their origin from the tibia.

φ χ χ : φ χ χ The peronei tertii, χ χ the tendon in the leg and foot.

ψ ω a b c : ψ ω a b c The long extenſors of the toes, ψ the tendon, ω a b c the four tendons into which it is divided running along the foot and ſmall toes.

d e f g h Theſe are only inſcribed on the firſt of the ſmall toes of the right foot, the ſmallneſs of the objects making it impoſſible upon the reſt; but they may all be eaſily underſtood from theſe, to which they are ſimilar, d the common tendon of the long and ſhort extenſors of the toes, inſerted in the bone of the ſecond phalanx, e the tendon running to the third phalanx proceeding from the ſhort extenſor of the toes: there is none ſuch belonging to the little toe, f the portion of the common tendon of the long and ſhort extenſor running to the third phalanx, g the common extremity of the two portions belonging to the third phalanx inſerted in it, h the aponeuroſis acceding to the tendon d, and proceeding partly from the capſular ligament of the joint of the toe with its metatarſus, partly from the interoſſeus muſcle of that ſide, partly from the lumbricalis muſcle, and partly from the ſide of the firſt phalanx.

i i i k : i i i k The tendons of the proper extenſors of the great toes, k the extremity inſerted in the laſt bone of the great toe.

l l l : l l l Branches of the tendons of the proper extenſors of the great toes found ſometimes.

m m Aponeuroſes, which the tendons of the proper extenſors of the great toes receive from the capſular ligaments of the joints of theſe toes with their metatarſal bones.

n o o p p p : n o o p p p The tibiales antici, o o their origin from the tibia, p p p the tendon.

q r s t : q r s t The ligaments by which the tendons are covered in the confines of the leg and back of the foot, r the ſuperior extremity, s fixed in the tibia, t the interior.

u w : u w The ligaments which bind down the tendons near the internal ancles, w their origin from the ancle.

x x The ligaments which bind down the tendons of the tibiales poſtici.

y y : y The tendons of the tibiales poſtici, partly inſerted in the navicular bones, partly extending to the greater cuneiform bones.

z z The heads which accede to the long flexors of the toes in the ſole of the foot, riſing from the heel bones.

IN THE EXTREMITY OF THE FEET, THE SHOULDERS, ARMS, &c.

A B C : A B C The abductors of the great toes, B the origin from the ſide of the heel bone, C the tendon.

D D The ſhort flexors of the great toes.

E E The ſhort flexors of the ſmall toes.

F F : F The tendons of the long flexors of the great toes.

G The tendon of the long flexor of the great toe, where it runs under the firſt phalanx of that toe contained in a ſheath and biſurcated.

I K : H I The ſhort extenſors of the toes, H the portion belonging to the great toe, I the portion running to the ſide of the firſt of the ſmall toes next the great one, found only in ſome ſubjects, K the portion belonging to the firſt of the ſmall toes.

L L The firſt interoſſei muſcles of the firſt ſmall toes.

M N O P Q : M N O Q The deltoïd muſcles, M the firſt portion of the firſt order whereof they conſiſt, N the firſt of the ſecond order, O P the third of the firſt, P its origin from the ſuperior proceſs of the ſcapula, Q the middle portion of the ſecond order.

R S : R S The coracobrachiales, R here united with the ſhort head of the biceps.

T T The long portions of the tricipites of the arm.

V V The ſhort parts of the ſame.

W X Y Z Γ: W X Y Z Γ The bicipites of the arm, W the long head, X the ſhort one, Y the common belly, Z the aponeuroſis which it gives to the tendinous vaginæ of the fore arm, cut off, Γ its tendon that is inſerted in the radius.

Δ Θ: Δ Θ The parts of the tricipites of the arm called brachiales externi, Θ the tendon that riſes from the ſurface of the brachialis externus, and reaches to the poſterior condyle of the arm bone.

ΛΛΛ: ΛΛΛ The brachiales interni.

Ξ The ſupinator brevis.

Π Σ: Π Σ The ſupinator longus of each arm, Σ the tendon.

Φ Φ The pronator teres of each arm.

Ψ Ω: Ψ Ω The radiales interni, Ω the tendon.

α β γ δ ε ε ε ε ε ε ε ε: α β the palmares longi, β the tendon, γ δ ε ε ε ε ε ε ε ε the aponeuroſis firſt ſlightly diſtinguiſhed into four portions, afterwards more ſo, and ſtrengthned with tranſverſe tendinous fibres, δ the portion which it gives to the ſhort abductor of the thumb, ε ε ε ε ε ε ε ε the extremities by which this aponeuroſis reaches to the roots of the fingers.

ζ η θ ι κ λ λ λ μ ζ ζ ζ the ſublimes, η θ the portion belonging to the middle finger, θ the tendon, ι κ the portion belonging to the third finger, κ the tendon, λ λ λ the portion belonging to the index, μ that of the little finger.

ν ξ ο The ulnaris internus, ξ the tendon, ο inſerted in the roundiſh bone of the carpus.

π ρ: π The long flexors of the thumbs, ρ the tendon.

σ The tendon of the profundus that goes to the index.

τ The pronator quadratus.

υ υ The ligaments under which run the tendons of the long abductors and the leſſer extenſors of the thumbs.

φ χ ψ ω b b: χ ω ω a b b The long abductors of the thumbs, χ the ſuperior part, ψ the inferior, ω ω the tendon of the ſuperior part, a the portion it gives to the ſhort abductor of the thumb, b b the tendon of the inferior part.

c d: c d The leſſer extenſors of the thumbs, d the tendon.

e The external armillary ligament.

f g g g g h The long radialis externus, g g g g h the tendon, h inſerted in the metacarpal bone of the index.

i The tendon of the other and leſſer long radialis externus.

k k l l l l The ſhorter radialis externus, l l l l the tendon.

m n o p q The common extenſor of the fingers, n o the portion belonging to the index, of which o is the tendon, p the tendon belonging to the middle finger, q that belonging to the ring finger.

r s The proper extenſor of the little finger, s the tendon.

IN THE LEFT HAND.

t the aponeuroſis by which the tendon of the index o, and of the middle finger p are united, and in like manner are united the tendons of the middle and ring fingers, and of the ring and little one: but neither upon theſe, nor upon the diviſions and conjunctions of the tendons of the extenſor communis and proper of the little finger are letters inſcribed on account of their ſmallneſs: and beſides the whole of them will be better underſtood from the firſt table of the back parts of the body.

u u The tendon of the indicator.

w The firſt interoſſeus muſcle of the index.

x The abductor of the index.

y The tendon of the greater extenſor of the thumb.

z The opponent muſcle of the thumb.

A B The common tendon of the greater and leſſer extenſor of the thumb, inſerted B in the laſt bone of the thumb.

C The aponeuroſis ſurrounding the capſular ligament of the joint of the thumb with its metacarpal bone, tied to that ligament, and joined to the common tendon of the extenſors of the thumb.

D The posterior tail of the short flexor of the thumb.

E The aponeurosis which the posterior tail of the short flexor of the thumb gives to the common tendon of the extensors of the thumb.

F G The adductor of the thumb, G its tendinous extremity, inserted in the first bone of the thumb.

H An aponeurosis, which rising partly from the first lumbricalis, partly from the abductor of the index, joins itself to the common tendon of the extensors of the index.

I The tendon of the first lumbricalis.

K L The common tendon of the indicator and extensor communis, running to the index, L its extremity inserted in the second bone.

M The tendon of the first lumbricalis, augmented by a portion received from the common tendon of the extensors of the index, running to the third bone of the index.

N The tendon of the posterior interosseus of the index which being augmented by a portion received from the common tendon of the extensors of the index, runs to the third bone of the index.

O The common tendinous extremity, inserted in the third bone of the index, this extremity is composed of the tendons M and N united into one.

P P The tendons of the common extensors of the fingers where they run along the back of the fingers with the aponeuroses which they receive.

Q The common tendon of the extensors of the little finger, where it runs along the back of that finger.

R The tendon common to the first interosseus of the middle finger and the second lumbricalis; which tendon being augmented by a portion received from the tendon of the extensor communis runs to the third bone.

S The tendon of the sublimis.

T The ligament that covers the tendon of the profundus, and also the extreme tails of the tendon of the sublimis.

V A tendon of the profundus.

The same S T V in the remaining three fingers: the letters are not inscribed on account of the smallness of the parts.

IN THE RIGHT HAND.

W W The ligament of the carpus, which together with the sinus of the carpus forms a canal, that contains and binds down the tendons of sublimis, profundus, and long flexnor of the thumb running from the fore arm to the fingers.

X The opponens of the thumb.

Y Z Γ Δ The short abductor of the thumb, Z its origin from the ligament of the carpus, Γ part of its tendinous extremity inserted in the first bone of the thumb, Δ a thin tendinous part that mounts upon the back of the thumb, and unites with the fore part of the tendons of the extensors of the thumb; and further is continued, along the exterior part of these tendons, to a similar aponeurosis of the short flexor of the thumb.

Θ The common tendon of the extensors of the thumb.

Λ Part of the short flexor of the thumb, which may be reckoned another short abductor of the thumb: its tendinous extremity inserted in the first bone of the thumb.

Ξ Ξ Π The tendon of the long flexor of the thumb, split as it were into two, Π the extremity that reaches to the last bone of the thumb.

Σ The oblique ligament by which the tendon of the long flexor of the thumb is fixed to the first bone of the thumb; at first one and then split into two tails.

Φ The posterior tail of the short flexor of the thumb.

Ψ The adductor of the thumb.

Ω The first lumbricalis.

a The first interosseus of the index.

b The abductor of the index inserted by its tendinous extremity in the first bone of the index.

c d e The abductor of the little finger, *d* its origin from the ligament of the carpus, *e* from the roundish bone of the carpus.

f The adductor of the metacarpal bone of the little finger.

g g The palmaris brevis.

h The small flexor of the little finger.

i The fourth lumbricalis.

k The third.

l The second.

m The first interosseus of the middle finger.

n The first of the ring finger.

o The first of the little finger.

p The common tendon of the small flexor and the abductor of the little finger.

q The common tendon of the fourth lumbricalis and the first interosseus of the little finger.

r The tendon of the posterior interosseus of the ring finger.

s The tendon common to the third lumbricalis and the first interosseus of the ring finger.

t The tendon of the posterior interosseus of the middle finger.

u The common tendon of the second lumbricalis and the first interosseus of the middle finger.

v The tendon of the posterior interosseus of the index.

w The tendon of the first lumbricalis.

x The tendon of the sublimis, on the part of which next the thumb is the tendon of the profundus, whereon no letter is put by reason of the smallness of the part.

y z The tendon of the profundus split as it were length-ways, *z* inserted in the third bone.

2, 2 The horns of the tendon of the sublimis.

3 The ligament that covers the tendon of the sublimis and profundus, where they run along the first phalanx, fixed to both margins of the first bone.

4, 4, 4 Three ligaments that retain the tendons of the sublimis and profundus at the joint of the finger with the metacarpus. They are thick, and by their middle thinner parts they are not only continued to one another, but also to the next ligament 3 of the same finger.

5 The ligament that covers the tendon of the profundus and the extreme tails of the tendon of the sublimis, about the middle of the second bone, fixed to both margins of the second bone.

The same, *x y z* 2, 2, 3, 4, 4, 4, 5 also in the other fingers, which easily appear tho' no letters are inscribed.

To the parts of the skeleton, that rise or are conspicuous between the muscles, I have put no marks; because they may be easily known from the first table of the skeleton, which figure is intirely the same, and the foundation of this, lying as it were or hid under it; for in order to construct this figure the muscles and other parts were placed upon that skeleton.

M

THE

EXPLICATION

OF THE

SECOND ANATOMICAL TABLE

OF

THE HUMAN MUSCLES.

The figure of this table is the back part of what is represented in the first. It likewise exhibits the whole muscular system, after the common integuments and tendinous vaginæ are removed: and moreover the ligaments belonging to the muscles; the ears, and part of the scrotum, and the naked parts of the skeleton.

IN THE HEAD, NECK, BACK, HIPS, AND THIGHS.

a b c : a b c d d e The epicranius or muscle of the scalp, *a b* the occipital muscle, *a* its tendinous beginning, *b* the fleshy part, *c d d* the aponeurosis between the occipital and frontal muscle, *d d* here the temporal muscle appears and rises behind it, *e* the membranous part by which the occipitals and their aponeuroses are united together; arising above the origin of the cucullares from the os occipitis.

f g The raiser of the ear, *f* its tendinous beginning where it rises from the epicranius, *g* its fleshy part.

h The frontalis.

i The orbicular of the eye-lids.

k The anterior of the ear.

l The lesser of the helix.

m n o The three retractors of the ear.

p q The masseter, *p* the posterior part of the interior portion, which is naked from the exterior portion, *q* the exterior portion.

r The greater zygomatic.

s The internal pterygoïd.

t The mylohyoïdeus.

t The latissimus colli.

u w : u w x The sternomastoïd with the cleidomastoïd united together, *w* the tendinous extremity, *x* inserted in the occipital bone.

y y The biventers of the neck, inserted in the occipital bone.

z z The splenii of the head.

A The levator scapulæ.

B C D E F G H H : B C D E F G H H The cucullares, B the fleshy part, C D E F the tendinous origin, C in this part rising from the occipital bone, D E F in this whole course it externally coheres with its fellow, below

below rising from all the spines of the back, the two inferior of the neck, and the ligament of the neck behind, E the large tendinous portion of the beginning at the lower part of the neck and upper part of the back, F another of the same kind in the lower angle, G the tendinous part of the extremity where it is inserted in the spine of the scapula not far from its basis, H H the tendinous part of the extremity inserted in the spine and superior process of the scapula.

I K : I K The infraspinati, K the origin from the base of the scapula.

L : L The greater rhomboïdei, inserted in the bases of the scapulæ.

M : M The sacrolumbales.

N : N The teres minor right and left.

O : O The teres major right and left.

P Q R R S T V : P Q R R S T V The latissimi dorsi, P the fleshy part, Q the broad tendon by which it begins, R R its origin from the spines of the loins and of the os sacrum, S its origin from the oblique processes, which lie at the side of the open of the os sacrum, T its cohesion with the great glutæus, V its origin from the crista of the ilium.

W X : W X The fleshy parts of the external oblique muscles of the abdomen, X X inserted in the crists of the ilium.

Y Z : Y Z a The glutæi medii, Z their origin from the os ilium, a the tendon.

β β The tensors of the vaginæ of the thighs.

γ δ δ : γ δ δ The glutæi magni, δ δ here it arises from the crista ilium and the sacrum, and coheres with the latissimus dorsi.

ε The levator of the anus. There is also a small part of the right one upon the right side.

Between ζ and ε the transverse of the perinæum.

η The external sphincter of the anus.

θ θ The great adductors of the thighs.

ι κ : ι κ The graciles, κ the tendon.

λ λ The sartorii.

μ μ The vasti interni.

ν ν ξ ο : ν ν ξ ο The semimembranosi, ξ the origin of the tendon from the fleshy part, ο the tendon.

π ρ : π ρ The semitendinosi, ρ the tendon.

σ τ τ υ φ χ : σ τ τ υ φ χ The bicipites of the legs, σ the longer head, τ τ the shorter head, υ φ χ the tendon; υ first arising from the surface of the fleshy part of the longer head, then augmented by the accession of the shorter φ, and by its extremity χ inserted in the superior head of the fibula.

ψ ω : ψ ω The vasti externi, ψ the tendininous surface.

IN THE LEGS, EXTREMITY OF THE FEET, AND SHOULDERS.

Γ ε : Γ ε The plautares

Δ : Δ The poplitæi.

Θ Λ Ξ : Θ Λ Ξ The peronei longi.

Λ Ξ Ξ Π Σ Σ Φ : Λ Ξ Ξ Π Σ Σ Φ The gemelli, Λ Ξ Ξ the exterior head, Ξ Ξ the tendinous surface, Π Σ Σ the interior head, Σ Σ the tendinous surface, Φ the tendon.

Ψ Ω : Ψ Ω The tendons of achilles, Ω Ω inserted in the heel bones.

a a b : *a a b* The solei, *b* the tendinous surface.

c c The tendons of the plautares.

d d The tendons of the tibiales postici.

Between the tendons *d* and Ψ in the left foot, and between *d* and the tendon of the plantaris in the right, lie the tendons of the long flexors of the toes.

e e The ligaments that bind down the tendons at the internal ancles, as they run near them.

f f The long flexors of the great toes.

g h i i k : *g h i k* The peronei breves, *h* the origin of the tendon from the fleshy part, *i i* the tendon, *k* inserted in the fifth metatarsal bone.

l m m m : *l m m m* The peronei longi, *m m m* the tendon.

n n The ligaments that bind down the tendons of the peronei longi and breves at the external ancles.

o o The ligaments proper to the peronei breves.

p p The ligaments proper to the peronei longi.

q q The ligaments by which the tendons in the confines of the legs and insteps are bound down.

r r The tendons of the long extensors of the toes.

s s The tendons of the peronei tertii, inserted in the metatarsal bones of the little toes.

t t The short extensors of the toes.

u w x y z : *u w x y z* The abductors of the little toes, *u* here covered with an aponeurosis, *w* the origin from the heel, *x* the aponeurosis by which the part is covered that is inserted in the metatarsal bone of the little toe, *y* the tendon of the abductor, inserted in the first bone of the little toe, *z* the aponeurosis acceding to that tendon of the long extensor of the toes, that belongs to the little toe.

α β : α The short flexors of the small toes, α the part inserted in the metatarsal bone of the little toe, β the part inserted in the first bone of the little toe by a tendinous extremity.

γ γ The tendon of the long flexor of the great toe, running between the sesamoïdal bones.

δ The abductor of the great toe.

ε The short flexor of the toes.

ζ η η θ ι κ λ μ ν : ζ η η θ ι κ λ μ The deltoïd muscles, ζ η η the second and posterior of the portions of the first order whereof that muscle consists, η η arising from the spine and superior process of the scapula, θ ι the posterior portion of the second order, ι arising from the superior process, κ λ the fourth portion of the first order, λ arising from the turn of the arm of the superior process, μ ν the middle portion of the second order, ν arising from the superior process.

ξ ο π ρ σ τ υ φ χ : ξ ο π ρ σ τ υ φ χ The tricipites of the arms, ξ the brevis, ο the longus, π the brachialis externus, ρ the common tendon of these three heads, σ the tendinous part made by the longus, and which is joined to the common tendon, τ of the right arm, the tendinous part made by the brachialis externus and joined to the common tendon: the τ of the left arm is placed at the rise of the tendinous from the fleshy part, υ the tendinous part arising from the surface of the brachialis externus, and reaching to the greater condyle of the os humeri, φ the common tendon inserted in the olecranon, χ the more slendor extremity of the same tendon, inserted in the fore part of the olecranon, and in the neighbouring part of the spine of the ulna.

ψ ψ The brachiales interni.

ω ω The supinatores longi.

IN THE FORE ARMS, AND RIGHT HAND.

A B C D D D D : A B C The longer radiales externi, B the origin from the lesser condyle of the arm-bone, C their conjunction and common origin with the common extensor of the fingers and the ulnaris externus, D D D D the tendon inserted in the metacarpal bone of the index.

E E The brachiales externi, arising from the roots of the lesser condyles.

F G : F The anconei, G the tendon, arising from the lesser condyle of the arm.

H : H I I I The shorter radiales externi, I I I the tendon.

K K The profundi, rising from the ulnæ.

L L The palmares longi.

M : M N O P Q The sublimes, N the portion going to the middle finger, O the part going to the index, P to the ring finger, Q to the little one.

R S T V : R S T V The ulnares interni, S T their origin, S the one rising from the greater condyle of the arm, and cohering with the common tendinous head of the muscles rising from that condyle, T the other rising from the olecranum, V the tendon inserted in the roundish bone of the carpus.

W X Y Y : W X Y Z The ulnares externi, X the origin conjoined with that of the common extensor of the fingers, Y Z the tendon, Z reaching to the fourth bone of the metacarpus. Between the tendon Z and the tendon *c*, on the back of the right hand, is a small tendon from this ulnaris externus reaching to the little finger.

a b c c *a b c* The proper extenſors of the little finger, *b* its origin conjoined with that of the common extenſor of the fingers, *c c* the tendon running along the back of the hand as it were ſlightly ſplit.

d: d e f f g h i k l m n o p p q r r s The common extenſors of the fingers, *e f f g h i k l m n o* the portion reaching to the ring finger, *f f* the tendon going to that finger along the back of the hand having inciſions, *g* a branch of this tendon which is afterwards ſplit into two, one of which *h* joins itſelf to the tendon *c* of the little finger, but is not always found; the other *i* is again ſplit into two, one of which *k* likewiſe accedes to the tendon *c* of the little finger, the other (between *k* and the lower *f*) accedes to the trunk *f* of the tendon of the ring finger, *l* a portion going from the tendon *f* and acceding to the tendon *c* of the little finger, *m* a tendinous portion by which the trunk of the tendon *f* going to the ring finger is joined with the tendon *c* of the little finger at the beginning of the fingers; which portion is compoſed of the portions *k* and *l*, and below theſe of the aponeuroſis going off from the tendon *f*, all united together, *n* the branch joined to the tendon *p* of the middle finger, not always found, *o* the tendinous portion by which the trunk of the tendon *f* going to the ring finger, is joined with the tendon *p* of the middle finger near the roots of the fingers; and this portion is compoſed of the tendon *n* united with the aponeuroſis that comes off from the trunk *f* of the tendon of the ring finger near the root of that finger, *p p* the tendon that goes to the middle finger, in which is a fiſſure where it runs along the hand, *q r r* the portion going to the index, *r r* the tendon, *s* the aponeuroſis which riſing from the tendon *p* of the middle finger, accedes to the tendon *r* of the index, and connects them together near the roots of the fingers.

t The tendon of the indicator.

u The tendon afterwards running along the index, compoſed of the tendon *t* of the indicator united with that tendon *r* of the common extenſor that goes to the index.

v w x y z z z z The tendons of the extenſors of the fingers where they run along the fingers joined with the tendons and aponeuroſes of the interoſſei, lumbricales, &c. *v* that of the index compoſed of the tendon *t* of the indicator conjoined with the tendon *r* from the extenſor communis, *w* that of the middle finger, *x* that of the ring finger; which two are from the common extenſor, *y* that of the little finger, which is compoſed of the tendon *c* of the proper extenſor of the little finger, conjoined with the portions *h* and *k l m* acceding from the common extenſor, and a portion from the ulnaris externus, *z z z z* the extremities of theſe tendons inſerted in the bones of the ſecond order.

α The aponeuroſis which reaches from the capſular ligament of the little finger with its metacarpus to the extenſor tendon *y*.

β γ The abductor of the little finger, *γ* the tendon.

δ ε The common tendon of the abductor and ſmall flexor of the little finger, *δ* conjoined with the tendon *y*, and augmented by a portion from it, *ε* it runs to the rhird bone.

ζ The aponeuroſis which accedes to the tendon *y*, its ſuperior part coming from the capſular ligament of the joint of this finger with its metacarpal bone, its inferior being produced from the tendon *η* of the interoſſeus of the little finger, with which tendon is joined the tendon of the fourth lumbricalis.

η The tendon of the interoſſeus of the little finger, to which is joined the tendon of the fourth lumbricalis.

θ ι The tendon common to theſe two muſcles, *θ* joined with the tendon *y* and being augmented by a portion from it, *ι* runs to the third bone.

κ The common extremity of the united tendons *ε ι* reaching to the third bone.

λ The aponeuroſis which accedes to the tendon *x*, its ſuperior part coming from the capſular ligament of the joint of this finger with its metacarpus; its inferior produced from the tendon *ν* of the poſterior interoſſeus of the little finger.

μ μ ν ξ ο The poſterior interoſſeus of the ring finger, *ν* the tendon that afterwards is joined *ξ* with the tendon *x*, and being augmented by a portion from it, *ο* runs to the third bone.

π The aponeuroſis that accedes to the tendon *x*, its ſuperior part coming from the capſular ligament of the joint of this finger with its metacarpus: the inferior being produced from the tendon *ρ* of the fore interoſſeus of the little finger, with which tendon is joined the tendon of the third lumbricalis.

ρ The tendon of the fore interoſſeus of the little finger, to which is joined the tendon of the third lum-

σ τ The tendon common to the fore interosseus of the little finger and the third lumbricalis, σ conjoined with the tendon *x*, and being augmented by a portion from it, τ it runs to the third bone.

υ The common extremity of the united tendons ο τ reaching to the third bone.

φ The aponeurosis that accedes to the tendon *w*, its superior part coming from the capsular ligament of the joint of this finger with its metacarpus: the inferior being produced from the tendon ψ of the posterior interosseus of this middle finger.

χ ψ ω Γ The posterior interosseus of the middle finger, ψ the tendon which is united ω with the tendon *w*, and being augmented by a portion from it, Γ runs to the third bone.

Δ The aponeurosis that accedes to the tendon *w*, its superior part coming from the capsular ligament of the joint of this finger with its metacarpus: the inferior being produced from the tendon Ξ of the fore interosseus of this middle finger, with which tendon is united that of the second lumbricalis.

Θ Θ Θ Λ Ξ The fore interosseus of the middle finger, Θ Θ Θ Λ the heads arising Θ Θ Θ from the metacarpal bone of the index, Λ from that of the middle finger, Ξ the tendon with which is united that of the second lumbricalis.

Π Σ The tendon common to the fore interosseus of the middle finger and the second lumbricalis, Π united with the tendon *w*, and being augmented by a portion from it, Σ runs to the third bone.

Φ The common extremity of the united tendons Γ Σ, reaching to the third bone.

Ψ The aponeurosis that accedes to the tendon *v*, its superior part coming from the capsular ligament of this joint with its metacarpus: its inferior being produced from the tendon 2 of the posterior interosseus of the index.

1, 2, 3, 4 The posterior interosseus of the index, 2 the tendon which is afterwards united 3 with the tendon *v*, and being augmented by a portion from it, runs 4 to the third bone.

5, 6 The aponeurosis that accedes to the tendon *v*, its superior part 5 being produced from the abductor of the index; and its inferior 6 from the first lumbricalis.

7 The tendon of the first lumbricalis, which afterwards is united 8 with the tendon *v*, and being augmented with a portion from it, runs 9 to the third bone.

10 The common extremity of the united tendons 4, 9 reaching to the third bone.

11 The fore interosseus of the index.

12 The abductor of the index.

13 The tendon of the great extensor of the thumb.

14, 15 The ligament that binds down the tendon of the ulnaris externus, arising from the radius between that ulnaris and the extensor of the little finger, ending at the tendon of the ulnaris internus, and here 15 united with the ligament 16.

16, 17, 18, 19 The exterior armillary ligament, arising 17 from the roundish bone, 18 from the triangular, 19 from the eminence of the radius that on the fore part terminates the sinus along which run the tendons of the radiales externi.

20 The ligament by which are bound down the tendons of the long abductor and lesser extensor of the thumb, in one part rising from the same eminence of the radius as the ligament 16.

21, 22, 23, 23 The long abductor of the thumb, 22 the tendon of the superior part, 23, 23 that of the inferior.

24, 25 The lesser extensor of the thumb, 25 the tendon.

26 The common extremity of the united tendons of the greater 13 and the lesser extensors 25 of the thumb reaching to the third bone.

27, 28 The aponeurosis that is joined to the common extremity 26 of the tendons of the extensors of the thumb: part of which aponeurosis 27 surrounds the capsular ligament of the joint of the thumb with its metacarpus, and is connected with that ligament; part of it 28 proceeds from the posterior tail of the short flexor of the thumb.

Between 27 and 29 the posterior tail of the short flexor of the thumb.

29, 30 The adductor of the thumb, 30 the tendinous extremity inserted in the first bone of the thumb.

IN THE EXTREME PART OF THE LEFT FORE ARM, AND IN THE LEFT HAND.

α β The exterior armillary ligament, β inserted in the roundish bone, and continued with the ligament γ δ.

γ δ The ligament that binds down the tendon of the ulnaris externus, δ ending at the tendon of the ulnaris internus.

ε The pronator quadratus.

ζ The ligament that makes a canal with the sinus of the carpus, by which the tendons are inclosed that proceed from the fore arm to the hand, viz these of the sublimis, profundus, and long flexor of the thumb.

η The portion of the tendon of the long abductor of the thumb, that it gives to the short abductor.

θ . κ The short abductor of the thumb, ι here it receives a portion from the aponeurosis of the palmaris longus, κ the tendinous extremity, with the aponeurosis that it gives to the tendon of the extensors of the thumb.

λ The part of the short flexor of the thumb, which may be looked upon as a second short abductor of it: inserted by its tendinous extremity in the first bone of the thumb.

μ Two ligaments, by which is bound down the tendon of the long flexor of the thumb: one of them higher at the joint of the thumb with its metacarpus: the other immediately below it, fixed to the margins of the first bone, at first single, afterwards bifurcated.

ν ν The tendon of the long flexor of the thumb, inserted in the last bone of the thumb.

ξ The posterior tail of the short flexor of the thumb, inserted in the first bone, and in the posterior sesamoïdal.

ο The first lumbricalis.

π The adductor of the thumb.

ρ The aponeurosis of the palmaris longus.

σ σ The palmaris brevis.

τ υ φ The abductor of the little finger, υ φ rising υ from the roundish bone, φ from the interior ligament of the carpus.

χ The small flexor of the little finger.

ψ The tendon common to the small flexor of the little finger, and the abductor of the same, united to the extensor tendon of that finger.

ω The extremity of the extensor tendon of the little finger reaching to the second bone.

Γ The tendon running to the third bone, composed of the tendon ψ, and of a portion of the extensor tendon of the little finger added to it.

Δ Δ The same tendons as ο τ υ, 4, 9, 10 in the right hand.

Θ The ligament that binds down the tendons of the sublimis and profundus, where they run along the first phalanx.

Λ Three ligaments that bind down the tendons of the sublimis and profundus at the joint of the finger with its metacarpus. I have inscribed a letter only on the middle one, on each side whereof lies one of the others.

Ξ The tendons of the sublimis and profundus.

Π The tendon of the profundus with one horn of the sublimis.

Σ The tendon of the profundus.

The same things are pointed out at Θ Λ : Ξ : Π : Σ in the other finger.

TAB. VI.

J. Brisbane M.D. delin. invenit edidit

J. Caldwall sculp.

THE EXPLICATION OF THE THIRD ANATOMICAL TABLE OF THE HUMAN MUSCLES.

As in the firſt table, ſo in this, the firſt order of the muſcles is repreſented, after the common integuments and tendinous vaginæ are removed; together likewiſe with certain ligaments belonging to them, and parts of the third ſkeleton, which is the baſis of this figure; alſo portions of other parts, to wit, of the noſe, ear, and genitals, which are not covered with muſcles.

IN THE HEAD, NECK, AND TRUNK.

a b c : a b c d e f The epicranius or muſcle of the ſcalp, *a b* the occipitalis, *a* its tendinous beginning, *b* the fleſhy part, *c d* the aponeuroſis between the frontal and occipital muſcles; *d* thro' which the temporal muſcle here riſing appears, *e* the membranous part, by which the occipitales and their aponeuroſes are joined together, riſing from the occipital bone above the origin of the cucullares, *f* the frontalis.

g h The raiſer of the ear, *g* its tendinous beginning, where it goes off from the epicranius, *h* the fleſhy part.

i The anterior of the ear.

k l m The three retractors of the ear.

n The greater of the helix.

o The leſſer of the helix.

p The tragicus.

q The antitragicus.

r s t t The orbicular of the eye-lids, *r* the part that goes round the circumference of the orbit, *s* the part added from the corrugator of the eye-brow, *t t* the part that covers the eye-lids.

u The compreſſor of the noſe.

v The naſal of the upper lip.

w x : w x The orbicular of the mouth, *x* the part on the red margin of the lip.

y z The greater zygomatic, *z* its origin from the jugal or cheek bone.

α β The depreſſor of the angle of the mouth, *β* its origin from the lower jaw.

γ The buccinator.

δ ε ζ η θ The maſſeter, *δ ε* the fore and exterior part, *ε* its origin from the jugal bone; from whence for a good ſpace it is tendinous externally, *ζ η θ* the poſterior part not covered by the former; *η θ* its origin, *η* from the jugal bone, *θ* from the zygomatic proceſs of the temporal bone.

ι κ The internal pterygoïd muſcle, *κ* inſerted in the lower jaw.

O

λ The ſtylohyoïdeus.

Between κ and λ the ſtylogloſſus. Compare Tab. X. fig. 2. *m*.

Immediately under the ſtylogloſſus the baſiogloſſus. Compare Tab. X. fig. 1. *k*. and fig. 2. *o*.

μ ν ν ξ ο π The latiſſimus colli, ν ν, &c. faſciculi or little bundles of fibres added to it in ſome ſubjects at the ſide of the neck, ξ ſcattered faſciculi vaniſhing on the cheek by which it ends, ο a faſciculus running along the fore part of the depreſſor of the angle of the mouth towards that angle.

π The lower jaw riſing under the latiſſimus colli.

ρ σ The ſternomaſtoïdeus and cleidomaſtoïdeus united together, σ their tendinous extremity.

τ The biventer of the neck inſerted by a tendinous extremity in the occipital bone.

υ The ſplenius of the head.

φ The ſplenius of the neck.

χ The ſcalenus medius.

κ ψ The raiſer of the ſcapula.

ω Γ Δ : ω Γ Θ Λ Λ Ξ The cucullares, Γ Δ the tendinous origin, Γ the part of it riſing from the occipital bone, Θ a larger part of its tendinous origin near the lower part of the neck and upper of the back, Λ Λ the tendinons part of the extremity, inſerted in the ſpine and ſuperior proceſs of the ſcapula, and in the neighbouring part of the clavicle, Ξ the tendinous part of its extremity where it is inſerted in the ſpine of the ſcapula not far from its baſis.

Π Σ Π The infraſpinati, Σ their origin from the baſis of the ſcapula.

Φ The teres minor.

Ψ The teres major.

Ω Ω A B C D E F The latiſſimi dorſi, Ω the fleſhy part, A the broad tendon by which it begins, B here it coheres with the great glutæus, C its origin from the criſta of the ilium, D E F the heads which riſe from the ribs, D from the eleventh, E from the tenth, F from the ninth.

G The ſerratus anticus.

H I The pectoralis, I the portion coming from the aponeuroſis of the external oblique of the abdomen.

K L M N O P Q R, &c. The great ſerratus, K the head that riſes from the third rib, L from the fourth, M from the fifth, N from the ſixth, O from the ſeventh, P from the eighth, Q from the ninth, R, &c. the place of the origin of the heads from the ribs.

S T U V W X Y Z *a a a a a b b b b b b c c d d d d e f f g h h h h i i i k* The external oblique muſcle of the abdomen, S the fleſhy part, T U V W X Y Z the heads whereof T riſes from the fifth rib, U from the ſixth, V from the ſeventh, W from the eighth, X from the ninth, Y from the tenth, Z from the eleventh, *a*, &c. the tendinous part of the origin of the heads, *b*, &c. the place of the origin of the heads from the ribs, *c c* the inſertion of the fleſhy part in the criſta of the ilium, *d d d d e f f g h h h h i i i k* the aponeuroſis, *e* inſerted in the criſta of the ilium, *f f* here the fleſhy part of the internal oblique riſes and appears under it, *g* here under the ſame and alſo under the aponeuroſis of the internal oblique riſes and appears the fleſhy part of the tranſverſe muſcle; *h h h h* and ſo here the fleſhy part of the rectus muſcle, *i i i* here through theſe aponeuroſes appear the tendinous lines of the rectus, *k* here under the ſame riſes and appears the pyramidalis.

l The cremaſter.

IN THE LEFT HAUNCH AND FOOT.

m The long adductor of the thigh.

n The pectineus.

o The great pſoas.

p The ſartorius.

q r s The tenſor of the vagina of the thigh, *r* its origin from the criſta of the ilium, *s* the extremity from which is cut the tendinous part that it gives to the vagina of the thigh.

t u u v The middle glutæus, *u u* its origin from the os ilium, *v* the tendon.

w x y The great glutæus, *x* here it rises from the crist of the ilium, and coheres with the latissimus dorsi, *y* the tendon.

z The semitendinosus.

α β β γ δ ε ζ The biceps of the leg, α the longer head, β β the shorter head, γ δ ε ζ the tendon, γ first rising from the surface of the fleshy part of the longer head, then augmented by an accession from the shorter δ, and its principal extremity ε inserted in the head of the fibula, and reaching to the tibia by a certain portion ζ.

η θ ι κ The vastus externus, θ the tendinous surface, ι the tendon, κ inserted in the patella.

λ μ ν ξ ο The rectus of the leg, μ the tendinous part of the origin, ν the tendon, ξ the place where it is inserted in the patella, ο the aponeurosis from the tendon of the rectus that runs along the fore part of the patella, and afterwards joins itself to the fore part of the ligament that reaches from the patella to the tibia.

π ρ The vastus internus, ρ the tendon.

σ τ υ The ligament going from the patella to the tibia, τ the place where it rises from the patella, υ under all this space it is inserted in the tibia.

φ χ ψ The exterior head of the gemelli, χ the tendinous surface, ψ the tendon.

ω Γ Δ The soleus, Γ rising from the superior head of the fibula, Δ the tendinous surface.

Θ Λ Ξ The tendon of achilles, Λ the interior part, Ξ here inserted in the heel bone.

Π The tendon of the plantaris.

Σ Φ Ψ Ω Ω Ω The peroneus longus, Φ its origin from the head of the fibula, Ψ Ω Ω Ω the tendon, Ψ here rising from the fleshy part.

A The proper ligament of the peroneus longus.

B B C C D D E The peroneus brevis, C C the origin of the tendon from the fleshy part, D D the tendon, E inserted in the fifth bone of the metatarsus.

F The ligament proper to the peroneus brevis.

G H I K L M N O P Q R The long extensor of the toes with the peroneus tertius, G the common fleshy part of the extensor and peronaus, H its origin from the tibia, I the peroneus tertius, K L M its tendon, K here rising from the fleshy part, M here inserted in the metatarsal bone of the little toe, N the tendon of the long extensor of the toes which is divided into four tendons, O P Q R running along the instep.

S S S T The tendon of the proper extensor of the great toe, T inserted in the last bone of that toe.

U V W W The tibialis anticus, V its origin from the tibia, W W the tendon.

X Y Z The ligament that covers the tendons in the confines of the leg and instep before, Y the superior horn, Z the inferior.

a a a b c d e The short extensor of the toes, *b c d e* its tendons, *b* that going to the great toe, *c* to the first of the small toes, *d* to the second, *e* to the third.

f g h i k l The common tendon of the long and short extensor of the toes, *f* the part produced by the long, *g* the part by the short: there is an intermediate mark of division, *h* the extremity inserted in the bone of the second order, *i* the portion of the common tendon going to the third bone, *k* the tendon from the other side running to the third bone and proceeding from the tendon of the short extensor, *l* the common extremity of the two portions going to the third bone, inserted in the third bone. The same in the other small toes of this foot, only the tendon of the long extensor runs along the upper part of the little toe, and produces both portions going to the third bone.

m The aponeurosis that is added from this side to the tendon of the short extensor of the toes. The same in the other toes: but in the small toe it is added to the tendon of the long extensor: in the great toe to the tendon of the proper extensor.

n The first interosseus muscle of the second toe.

o The thicker head of the second interosseus of the second toe.

p The thicker head of the second interosseus of the third toe.

q The thicker head of the second interosseus of the fourth toe.

r s t u v The abductor of the little toe, *s* here covered with an aponeurosis, *t* its origin from the heel bone, *u* part of the aponeurosis with which it is covered inserted in the metatarsal bone of the little toe, v the tendon of the abductor inserted in the first bone of the little toe.

w x The short flexor of the little toe, w the part inserted in the metatarsal bone of the little toe, x the part inserted by a tendinous extremity in the first bone of the little toe.

y z The tendons of the long and short flexors of the toes, seen also in the next toe.

IN THE RIGHT FOOT.

A B C The rectus of the leg, B the tendon inserted in the patella, C the aponeurosis from the tendon running over the patella, and joining itself to the fore part of the ligament that joins the patella to the tibia.

D E F G The ligament that joins the patella to the tibia, E the part under which it rises from the patella, F under all this space it is inserted in the tibia, G its interior part.

H I The vastus internus, I the extremity of its tendon inserted in the side of the patella.

K L M The sartorius, L the tendon, M inserted in the tibia.

N O The gracilis, O its tendon.

P Q R S The semimembranosus, Q R the tendon, Q rising here from the fleshy part, S the anterior aponeurosis inserted in the internal margin of the tibia.

T U The semitendinosus, U the tendon.

V W X The interior head of the gemelli, W the tendinous surface.

X The tendon, it joins to the exterior part of the tendon of the soleus.

Y The tendon of achilles, Z inserted in the heel bone.

Γ Δ Θ Θ The soleus, Δ the tendinous surface, Θ Θ rising from the tibia.

Λ Ξ The tendon of the plantaris, Ξ inserted in the heel bone.

Π Σ The long flexor of the great toe, Σ the tendon.

Φ The ligament that binds down the tendon of the long flexor of the great toe.

Ψ Ω Ω α The long flexor of the toes, Ω Ω its origin from the tibia, α the tendon rising from the fleshy part.

β β γ The tendon of the tibialis posticus, γ its extremity inserted in the tuberosity of the navicular bone.

δ ε ε The ligament that covers the tendon of the long flexor of the toes, and that of the tibialis posticus, ε ε here fixed to the internal ancle.

ζ The ligament that binds down the tendon of the tibialis posticus.

η θ θ θ The tibialis anticus, θ θ θ its tendon.

ι κ The superior horn of the ligament inserted in the tibia by which the tendons are bound down, in the confines of the leg and instep, inserted in the tibia, κ the inferior horn of the same ligament.

λ λ μ The tendon of the proper extensor of the great toe, μ inserted in the last bone of the great toe.

ν A branch of the tendon of the proper extensor of the great toe, inserted in the first bone of that toe and found in some subjects.

ξ The aponeurosis added to the tendon of the proper extensor of the great toe.

ο Upon these toes the common tendons of the extensors.

π π The two horns of the ligament, by which the tendon of the long flexor of the great toe is bound down here.

ρ The tendon of the long flexor of the great toe running under that toe.

σ ς ς τ The abductor of the great toe, ς its origin from the heel bone, ς its tendon, τ inserted in the first bone of the great toe.

υ υ φ The short flexor of the great toe, φ here joining itself to the tendon of the abductor.

χ ψ The short flexor of the toes, χ its origin from the heel bone.

ω The head that is added to the long flexor of the toes in the sole of the foot rising from the heel bone.

IN THE LEFT ARM AND HAND.

A B B C D E F G H I K L M N O The deltoïd muſcle, A B B the ſecond and poſterior portion of the firſt order of which that muſcle conſiſts, B B riſing from the ſpine and ſuperior proceſs of the ſcapula, C D the poſterior portion of the ſecond order, D riſing from the ſuperior proceſs, E F the fourth portion of the firſt order, F riſing from the turn of the ſuperior proceſs, G H the middle portion of the ſecond order, H riſing from the ſuperior proceſs, I K the third portion of the firſt order, K riſing from the ſuperior proceſs, L M the firſt portion of the ſecond order, M riſing from the ſuperior proceſs, N O the firſt and fore portion of the firſt order, O its origin from the ſuperior proceſs.

P Q R The biceps, Q its tendon, R the aponeuroſis cut off.

S The brachialis internus.

T U V W X The triceps, T the part of it called longus, U the part called brevis, V the tendon, W inſerted in the olecranon, X the more ſlender end of the tendon inſerted in the fore part of the olecranon, and in the neighbouring part of the ſpine of the ulna.

Y The ulnaris internus.

Z *a* The ſupinator longus, *a* the tendon.

b The pronator teres.

c d The radialis internus, *d* the tendon.

e e The ſublimis.

f f The long flexor of the thumb.

g The tendon of the ſecond radialis externus longior.

h i i i k The radialis externus longior, *i i i k* the tendon, *k* inſerted in the metacarpal bone of the index.

l m m m The radialis externus brevior, *m m m* the tendon.

n o The ulnaris externus, *o* the tendon.

p q r s t The common extenſor of the fingers, *q* the tendon belonging to the index, *r* that belonging to the middle finger: which two tendons not far from the fingers are joined by the portion *s*, going from the tendon of the middle finger to that of the index, *t* the tendon going to the third finger.

u The tendon of the indicator.

v w The proper extenſor of the little finger, *w* the tendon.

x y y z z The long abductor of the thumb, *y y* the tendon of the ſuperior portion which being divided at the extremity ſends one part to the ſhort abductor, and the other to the metacarpal bone of the thumb, *z z* the tendon of the inferior portion.

α β The leſſer extenſor of the thumb, *β* the tendon.

γ γ The tendon of the greater extenſor of the thumb.

δ ε The common tendon of the greater and leſſer extenſor of the thumb, *ε* reaching to the laſt bone of the thumb.

ζ ζ The exterior armillary ligament.

η The ligament that binds down the tendons of the long abductor and leſſer extenſor of the thumb.

θ The interior ligament of the carpus.

ι κ The ſhort abductor of the thumb, *κ* the aponeuroſis which it gives to the common tendon of the extenſor of the thumb.

λ The opponens of the thumb, inſerted in the external margin of the metacarpal bone of the thumb.

μ The aponeuroſis that joins itſelf to the tendon of the greater extenſor of the thumb, partly riſing from the capſular ligament of the joint of the thumb with its metacarpus, partly from the ſhort flexor of the thumb.

ν ξ The adductor of the thumb, *ξ* inſerted by a tendinous extremity in the firſt bone of the thumb.

ο The fore interoſſeus of the index, riſing from its metacarpal bone.

π The abductor or the index.

ρ The firſt lumbricalis.

σ The aponeuroſis that partly riſes from the abductor of the index, partly from its lumbricalis, and is joined to the tendon of the extenſors of the index.

ς τ The common tendon of the extenſors of the index, *τ* its extremity inſerted in the ſecond of the index.

τ υ The tendon of the first lumbricalis, which being augmented by a portion received from the common tendon of the extensors, υ runs to the third bone of the index.

φ The tendon of the second interosseus of the index, together with the portion it has received from the common tendon of the extensors, running to the third bone of the index.

χ The common extremity of the tendons υ and φ, belonging to the third bone of the index and inserted in it.

ψ The tendon common to the second lumbricalis and the first interosseus of the middle finger: which tendon being augmented by a portion received from the tendon of the extensor of the middle finger, runs to the third bone of that finger, in the end joined into a common extremity with a like tendon coming from the other side of that finger, which is inserted in the third bone of the same finger.

ω The tendon common to the fourth lumbricalis and the first interosseus of the little finger: which tendon being augmented by a portion received from the extensor tendon of that finger, runs to the third bone of the same.

Along the internal parts of the fingers run the tendons of the sublimis and profundus, bound down by their ligaments, but they are more clearly seen in the first table of the muscles.

IN THE RIGHT ARM.

α β γ δ ε The triceps of the arm, α the part called longus, β the part called brachialis externus, γ the tendon of the triceps, δ inserted in the olecranon, ε the thin tendon, rising from the surface of the brachialis externus, and reaching to the upper part of the greater condyle of the arm bone.

ζ The brachialis internus.

η θ ι The biceps of the arm, θ the aponeurosis cut off, near ι the tendon.

κ The supinator longus.

λ The pronator teres.

μ The radialis internus.

ν ξ The palmaris longus, immediately below ξ the beginning of the tendon.

ο The sublimis.

π ρ σ The ulnaris internus, ρ one of its origins, from the greater condyle of the arm bone, σ the other from the olecranon.

τ The ulnaris externus.

IN THE RIGHT HAND.

a b The short abductor of the thumb, *b* the aponeurosis, going off from its tendon and joining itself to the common tendon of the extensors of the thumb.

c Part of the short flexor of the thumb, which may be esteemed a second short abductor of it, inserted by its tendinous extremity in the first bone of the thumb.

d That part of the short flexor of the thumb that is inserted in the sesamoïdal bone next to the index, and to the neighbouring part of first bone of the thumb.

e The adductor of the thumb.

f f The tendon of the long flexor of the thumb.

g Two ligaments that bind down the tendon of the long flexor of the thumb, μ in the left hand Tab. V. of Albinus.

h The palmaris brevis.

i The short flexor of the little finger.

k The abductor of the little finger.

l The adductor of the fourth metacarpal bone, inserted in that bone.

Letters are not inscribed upon the tendons and aponeuroses that run along the back of the hand and fingers, as they may be easily understood from the second muscular table, which is the fifth, as this third is the ninth of Albinus.

THE

ANATOMY OF CELSUS,

AND

PHYSIOLOGY OF CICERO.

P R E F A C E.

THE reasons for adding the following translations from Celsus and Cicero may be seen in the general preface; and it were to be wished that the modern writers on these subjects, would endeavour to imitate such excellent models: for tho' the ancients were inferior to the moderns in the accurate knowledge of nature, in so far as it depends on minute observation and experiment, yet they were the original fountains of the great principles in almost all the arts and sciences, and treat of them, not with obscure diligence, but in a noble and masterly manner, adorned with all the charms of eloquence; and none more remarkably than the two authors of which a specimen is here given. The Græcian philosophy was a favourite study of Cicero, and he was the first that introduced it to his country in a Roman dress; for when banished from the forum and public affairs, by the disorders of the commonwealth, philosophy, the precepts of which he had made the rule of his life, was his consolation in the calamities of the public, and in his private distresses; and in this treatise on the nature of the Gods, the existence, attributes, and providence of the SUPREME BEING are proved, from that best and clearest of all evidence, the excellency of the works of nature; and this he concludes, with the beautiful survey of man here presented to the reader. Celsus lived in the age posterior to Cicero. He was a noble Roman, as appears not only from his name, but from the spirit, the manner, and generous sentiments of his writings: he wrote in a masterly strain upon many important subjects, for which we have not only the testimony of his admirers, but even that of Quintilian, a writer several degrees inferior to Celsus, and who seems to have envied the brightness of his fame. Of the writings of Celsus that upon medicine only remains; a work full of the genius of antiquity, and a complete model to medical writers, and above all others fittest to give the curious a true idea of that art. He comprehends, in a small volume, a complete system of medicine, as it ſtood in the Auguſtan age, with the utmoſt judgment, perſpicuity, and elegance, and with a freedom and eaſe, hardly to be found in the moſt diffuſe writers. Of anatomy Celſus treats in a curſory way, and in ſo far only as is needful to illuſtrate medicine. His moſt complete part, upon this ſubject, is the chapter upon the bones, before he comes to treat of their diſeaſes; alſo his chapter on the viſcera of the thorax and abdomen, before he treats of the diſeaſes that affect the particular parts ſeated there. He has likewiſe two ſhort chapters of the anatomy of the eye, and of the teſticles; which are more obſcure and imperfect. To theſe when we add the general maxims upon the uſe of anatomy in medicine, to be found in his general preface, we then have a view of the whole anatomy of Celſus; in which, if we except ſome trivial errors, and allow for the evident corruptions of the text, a more maſterly deſcription in ſo ſmall a compaſs is no where to be found; and I think, in general, evidently taken from the human body. Nor can it be ſuppoſed, that any man could deſcribe the bones, and even the viſcera, as he has done, without having ſeen and ſtudied them, not only with care, but, which is much rarer, with taſte and judgment; and the ſame may be ſaid of the particulars contained in almoſt every part of his ſyſtem of medicine, in which he is ſo far ſuperior to moſt medical writers. A few notes and obſervations on the anatomical part of Celſus are added at the end of the text, which, tho' I have tranſlated with all the care I was able, I am afraid the admirers of this fine author will find but a very imperfect picture of the original.

THE ANATOMY OF CELSUS.

Book IV. Chap. I.

Of the situation of the inward parts of the human body.

HITHERTO we have treated of thofe kinds of difeafes which belong to the whole body, fo that certain feats cannot be affigned them; now I fhall treat of thefe which belong to particular parts. But the difeafes and cures of all the inward parts will be more eafily known, if I firft mention their feats in a brief manner, giving a flight view of the parts, in fo far as is neceffary to the healing art.

The head therefore, and thefe parts that are within the mouth, are not only bounded by the tongue and palate, but alfo by all that our eyes can there difcover. In the right and left fides about the throat, the large veins called fphagitides, and likewife the arteries which they call carotides, proceeding upwards, reach beyond the ears. And on the neck itfelf are placed glands, which fometimes fwell with pain. From thence two paffages begin; the one they call afpera arteria or wind pipe, the other ftomachus or gullet. The artery is exterior, and leads to the lungs; the gullet is interior, and leads to the ftomach: the firft receives the breath, the other the food. As thefe are diftinct paffages, where they meet, there is a little tongue upon the artery, juft in the fauces, which is raifed up when we breath, but when we eat or drink is depreffed, and fhuts the artery. The artery itfelf, being hard and cartilaginous, is protuberant in the throat, but more funk in other parts. It confifts of certain circles, made in the form of thefe vertebræ which are in the back bone; fo that on the outfide they are unequal, on the infide they are fmooth, like the gullet; and defcending is joined to the lungs in the præcordia. The lungs are of a fpongy nature, and therefore fit to contain air; they are connected behind to the back bone, and are divided into two lobes like an ox's hoof. To thefe the heart is annexed, of a mufcular fubftance, placed in the breaft under the left mamma, and has as it were two ventricles. Under the heart and lungs there is a tranfverfe divifion, confifting of a ftrong membrane, which divides the præcordia or thorax from the lower belly; it is tendinous equally all over, and has many blood veffels running upon it, and divides not only the inteftines, but alfo the liver and fpleen from the upper parts. Thefe bowels are placed near, but under it, on the right and left fides. The liver on the right fide, under the præcordia, arifing from the diaphragm itfelf, is concave on the infide, and on the outfide convex; and ftretching out, rides for fome way upon the ftomach, and is divided into four lobes. Upon the inferior part of the liver, the gall-

bladder

bladder is joined; but the ſpleen on the left ſide is not attached to the ſeptum, but to the inteſtine; it is of a ſoft and rare texture, and of a moderate length and thickneſs, and being chiefly concealed under the ribs, ſtretches ſomewhat from thence into the abdomen: all theſe parts are connected together: but the kidneys are divided; they are ſeated in the loins, about the loweſt ribs, on one ſide round, on the other hollowed; they are full of blood veſſels, have ventricles, and are covered with coats: ſuch is the ſituation of the viſcera: but the ſtomachus or gullet, which is the beginning of the inteſtines, takes its riſe nervous from the ſeventh vertebra of the ſpine, and is joined with the ſtomach about the præcordia. The ſtomach, which is the receptacle of the food, conſiſts of two coats, and is placed between the ſpleen and the liver, both of theſe riding a little over it; and there are alſo ſlender membranes, by which theſe three are joined together, and they are connected to that tranſverſe ſeptum, which I juſt now deſcribed: thence the lower part of the ſtomach, turned a little to the right ſide, is contracted into the ſize of the firſt inteſtine; this joining the Greeks call pylorus, becauſe like a door, it emits into the lower parts what we are to excrete. From this the inteſtine called jejunum begins, which is not much involved; it is ſo named becauſe it never retains what it receives, but immediately tranſmits it into the lower parts: after this follows the ſmall inteſtine, greatly convoluted into plaits, each convolution of which is connected to the more inferior by little membranes, and the whole leaning towards the right ſide, and ending oppoſite to the right haunch bone, fills chiefly the upper parts of the abdomen. Then this inteſtine is joined tranſverſly with the other thicker one, which beginning on the right ſide of the abdomen, is open and ſtretched out towards the left, but not towards the right, and is therefore called cæcum; but that which is open and pervious, is ſpread wide and in arches; and being leſs nervous than the upper inteſtines, folded here and there on each ſide, yet poſſeſſing chiefly the left and inferior parts, touches the liver and ſtomach, and there is connected with certain membranes coming from the left kidney; and from thence making a turn towards the right ſide, is directed downwards to where it ſends out the excrements, and therefore it is called in that place the ſtraight gut. The omentum covers all theſe, on the under part ſmooth and conſtricted, on the upper part more ſoft; and on this fat grows, which is devoid of ſenſe like the brain and marrow. But from each kidney proceeds a vein of a white colour, the Greeks call them ureters, becauſe they imagine that the urine deſcending thereby diſtils into the bladder. This, in its great cavity nervous and double, with a full and fleſhy neck, is joined by veins with the inteſtine, and with that bone which is under the pubes; itſelf is more looſe and free, placed differently in men and in women: for in men it is next the ſtraight gut, and rather inclining towards the left ſide; in women it is placed upon their genitals, and riſing above it, is ſupported by the womb itſelf. Beſides, in males the paſſage of the urine, being longer and narrower, deſcends from the neck of the bladder to the penis; in women it is ſhorter and larger, and ſhows itſelf above the neck of the womb: the womb in virgins is very ſmall, and even in women, if they are not pregnant, is not much larger than may be comprehended in the hand. It begins with a ſtraight and continued neck, which is called the canal, oppoſite to the middle of the abdomen; from whence it is a little turned towards the right coxa, and then ſtretching itſelf upon the ſtraight gut, connects its lateral parts to the ilia of the woman. Theſe are ſituated between the coxæ and the pubes at the bottom of the abdomen, from which and from the pubes the abdomen proceeds upwards to the præcordia, on the outſide incloſed by the ſkin which we ſee, and on the inſide by a ſmooth membrane that lies next the omentum, and is called peritoneum by the Greeks.

CELSUS

CELSUS is very full on the diseases of the eyes, which he introduces with the following short but elegant exordium:

What I have been treating of is of less importance; but our eyes are subject to various and great disorders; and as they contribute so much to the uses and pleasures of life, the greatest care should be bestowed on them.

BOOK VII. CHAP. VII. SECT. 13.

OF THE NATURE OF THE EYES.

I HAVE mentioned the suffusion in another place, because when recent it is often cured by medicines; but when it is of long standing, it demands the assistance of the hand, and it may be reckoned among the nicest operations. But before I treat of this, I shall first in a few words shew the nature of the eye itself, the knowledge of which, tho' it belongs to many articles, yet is chiefly requisite in this place. The eye has two principal coats, of which the superior is called by the Greeks the ceratoeïdes, or the horn-like; this is pretty thick in the white part of the eye, but thinner near the pupil. To this is conjoined an inferior coat, pierced in the middle by a hole of a moderate size, which makes the pupil, and there it is thinner; but in the farther parts this likewise becomes thicker, it is called choroeïdes by the Greeks. These two coats having surrounded the internal parts or contents of the eye, at last unite behind them, and becoming thin and gathered together, pass thorough the hole between the bones, to the membrane of the brain, and are attached to it. But at the pupil there is an empty space next these coats, and again below them there is a very thin membrane, which Herophilus called arachnoeïdes; this hangs down in the middle, and in this cavity something is contained, which from its likeness to glass the Greeks called naloïdes; this is neither liquid nor solid, but as it were a concreted humour, from the color of which, that of the pupil is either black or blewish, tho' the whole outer coat be white; but a little membrane covers it, coming from the interior parts; near these is a drop of humour, like the white of an egg, from which we have the power of vision, the Greeks called it crystalloïdes.

CELSUS introduces the diseases of the private parts with a short preface, that shews a sense of modesty not common even in the politest authors of the Augustan age, and for that reason the more to be admired:

The next in order, *says Celsus,* are the diseases of the private parts, the Greek names of which are more decent, and already familiarized by use, as they occur in the common books and language of physicians: with us the terms are shocking, nor are they recommended by the authority of any one that writes or speaks with modesty, so that it is difficult to explain these things, as the rules of art and of modesty at the same time demand. But this must not deter me from writing upon the subject, not only that I may complete my system of the healing art, but because the cures of these parts should be more generally known which we are less willing to expose to others.

Book VII. Chap. XVIII.

Of the Nature of the Testicles.

I COME now to thefe difeafes which arife in the private parts about the tefticles; and in order the better to explain them, we muft firft in a few words defcribe the nature of the parts. The tefticles therefore have fomething that refembles a medullary fubftance, for they do not emit blood, and are quite infenfible; but the coats that contain them are fubject to pain in ftrokes and inflamations. They hang down from the groins by nerves or cords, which the Greeks call cremafters, with each of which a vein and an artery defcends. They are covered by a thin nervous coat without blood, and white, which the Greeks call elutroeides. Over this is a ftronger coat, which adheres firmly on the lower part to the inner one; the Greeks call it dartos. Then many little membranes furround the veins, arteries, and thefe nerves, and the like are alfo found very thin and fmall on the upper part between the two coats. Such are the coverings and defence proper to each tefticle. But there is an outer vifible bag common to both, and to all the internal parts; the Greeks call it ofcheon, we the fcrotum. This is flightly attached to the middle of the coats on the lower part, and above only furrounds them.

Book VIII. Chap. I.

Of the Position and Figure of the Bones of the Whole Human Body.

THE part remains relating to the bones: to underftand which, it is neceffary to defcribe their pofition and figures. I fhall begin with the fkull, which is inwardly concave, outwardly convex; on each fide fmooth, both where it covers the membrane of the brain, and where it is covered by the hairy fcalp. About the hind head and temples it confifts of one plate only, but from the forehead to the vertex it confifts of two. The bones are harder outwardly, but fofter or more fpungy in the inward parts by which they are connected; and fmall veffels run between them, which probably fupply them with nourifhment.

It is rare to find a fkull folid without futures, except in very hot climates, and fuch heads are ftronger, and lefs fubject to pain: and in general, the fewer the futures are, fo much better health does the head enjoy. Neither the number nor the places of the futures are fixed or certain. Yet two above the ears almoft conftantly appear, and feparate the temples from the upper parts of the head. A third, ftretching to the ears from the vertex, divides the hind head from the upper part. A fourth, from the fame vertex, paffes along the middle of the head to the forehead, and there fometimes ends where the hair begins; at other times cutting the forehead, it lands between the eye-brows. All thefe futures are indented into each other, except the tranfverfe ones above the ears, which are gradually thinned along their edges, and fo placed that the lower bones are gently

feated

feated upon the upper ones. But the face has the largeft future of all, beginning at one temple it proceeds through the eyes and nofe, tranfverfely to the other, from which two fhort futures point downwards, from the inner angles of the eyes. The cheek bones too have each a tranfverfe future, on the upper part; and from the middle of the noftrils, or roots of the upper teeth, a future proceeds along the middle of the palate, and another tranfverfe one cuts the fame palate: fuch are the futures in moft fubjects.

Of the holes belonging to the head, thefe of the eyes are largeft; the next are thefe of the noftrils; and laftly of the ears. Thefe of the eyes tend fimply and directly towards the brain; the two holes of the nofe are divided by a middle bone. The nofe begins with bone at the eye-brows and angles of the eyes, for about a third part; from thence it is changed into cartilage, and as it approaches towards the mouth, it is even foftened into flefh. But each hole of the nofe, tho' to a certain depth it is fimple and one, becomes at length divided into two paffages, one whereof going to the fauces, receives and fends forth the breath; the other tends to the brain, the extremity of which is divided into fmall holes, whereby we have the fenfe of fmell. In the ear alfo the paffage is at firft ftraight and fimple, but as it proceeds becomes winding, and near the brain is divided into many fmall paffages, by which we have the power of hearing. The thickeft bone of the head is that behind the ear, for which reafon probably no hair grows on that part. Under the temporal mufcles, the middle bone is placed inclining outwards. Near thefe are as it were two fmall finuofities; and above them that bone ends, which, tending tranfverfely from the cheeks, is fupported by the bones below; it may be called the yoke or jugal bone, from the fame fimilitude that the Greeks called it zugodes. The lower jaw is a foft or cellular bone and fingle; the middle and lower part of it forms the chin, from which it extends on each fide to the temples, and this alone enjoys motion; for the cheek bones, with the whole of that which contains the upper teeth, are immoveable. The extremities of the lower jaw ftand up like two horns; the one procefs, thicker below, becomes thin at the top, and rifing higher than the other, goes under the os jugale, and is there fixed to the mufcles of the temples. The other is fhorter and more round, and being fitted to that finuofity which is near the hole of the ear, by moving itfelf every way like a hinge, is the caufe of all the motions of the lower jaw.

The teeth are harder than bone. Of thefe, part belong to the lower jaw, and part to the fuperior bone of the cheek. The firft four teeth, becaufe they cut, are called tomicoi by the Greeks; next thefe are the four canini; beyond which on each fide there are commonly five maxillares, except in thefe perfons who have not got their genuini or lateft teeth. The fore teeth fpring from one fingle root, the maxillary have two, and fometimes three or even four roots; and the longeft roots commonly produce the fhorteft teeth: the ftraight teeth have ftraight roots, the crooked have bent ones. From this root in children a new tooth comes forth, which commonly expels the former, but fometimes appears above or below it.

After the head comes the fpine, which confifts of four and twenty vertebræ; feven in the neck, twelve at the ribs, and the other five below the ribs. The vertebræ are fhort and rounded, and fend out two proceffes from each fide; they are perforated in the middle, by which the fpinal marrow defcends from the brain: on the fides likewife, between the two proceffes, they are flightly hollowed and pervious, by which are conveyed from the membranes of the brain fimilar little membranes; and all the vertebræ, except the three higheft, have fmall finuffes funk in their proceffes above,

and

and on the lower part fend other proceffes downwards. Thus the higheft vertebra immediately fuftains the head, whofe fmall proceffes are received by the two finuffes of the vertebra; the head for this purpofe being provided with two rough proceffes inclined upwards and downwards. The fecond vertebra is inferted into the inferior part of the firft, and its circumference is finifhed above in a leffer circle; therefore the firft vertebra comprehending the fecond, allows the head a motion towards each fide. The third receives the fecond in the fame manner, by which the neck has a free and eafy motion. But the neck could not fuftain the head, if ftraight and ftrong ligaments did not affift it on each fide; thefe the Greeks call tenontes, and in every motion one of thefe is always ftretched, fo as not to allow the parts above to give way. Now the third vertebra fends out tubercles which are inferted in the one below, and all the reft are infinuated into thefe under them, by means of proceffes that point downwards; and they receive the vertebræ above them into finuffes which they have on each fide, and are bound together by means of ligaments and much cartilage; and thus by one moderate turn properly given, a man can at one time ftand erect, and at other times bend his body in all the neceffary offices of life.

Below the neck, the firft rib is placed about the fame height with the fhoulders; and after it, fix lower ribs extend to the bottom of the breaft; thefe at their beginning are rounded as it were with little heads, which are inferted in the tranfverfe proceffes of the vertebræ, flightly hollowed for that purpofe; from thence they become broad, and being twifted outward they by degrees degenerate into cartilage; and at that part being again flightly bent inward, they are joined to the pectoral bone. This begins ftrong and hard near the fauces, is lunated on each fide, and ends at the præcordia, foften'd there into a cartilage. Under the former ribs are five, which the Greeks call nothai, being fhort and more flender, and like the others gradually changed into cartilage; thefe have their extremities in the abdominal parts, and the loweft of them confifts chiefly of cartilage.

Again, immediately below the neck, two broad bones, one on each fide, tend to the fhoulders; the Latins call them fcoptula aperta, the Greeks omoplatæ; thefe have a finus at the upper part, from thence are triangular, and fpreading gradually reach towards the fpine, the broadeft parts being always the blunteft: thefe bones are cartilaginous at bottom, and hang as it were loofe on the back parts, for except at the top they are joined to no bone, but are there fixed by ftrong ligaments and mufcles. Near the firft rib, fomewhat more internally than its middle, this bone has an excrefence, at that part thin, but proceeding nearer the broad bone of the fcapula it becomes thicker and broader, and is bent a little outward; near which, fwelling fomewhat, on another part of its fummit, it fupports the collar bone.

This is crooked, and may be reckoned among the hardeft bones, with one head refting on the part I mentioned, and with the other on a little finuofity of the pectoral bone; and it yields fomewhat along with the motions of the arm, its head being connected below with the fcapula by ligaments and cartilage.

Thence begins the arm bone, fwelling with heads at both extremities; and there it is foft without marrow, and covered with cartilage; the middle or body is cylindrical, hard, and full of marrow, gently convex on the fore and internal part, and outwardly concave: by the fore part I mean the plain of the breaft; by the parts behind, that of the fcapulæ: the internal is that which is next the fide, and the external that which is removed from it. Thefe diftinctions I fhall all along apply to every member.

member. The upper head of the arm is rounder than the other bones of which I have ſpoken, and is inſerted in the broad bones of the ſcapulæ at the top thereof, and is tied outwardly in that ſituation chiefly by ligaments. Below it has two proceſſes, between which at the extreme parts there is a large ſinuoſity.

Theſe afford a place to the fore-arm, which conſiſts of two bones, the radius, which the Greeks call cercis, is the ſuperior and ſhorteſt; it is more ſlender at the upper part, and receives into its round and ſlightly hollowed head a ſmall tubercle of the humerus, which is fixed there by ligaments and cartilage. The cubit or ulna is inferior and longer, being thickeſt at top, and there by two proceſſes is inſerted into the ſinuoſity of the humerus, which I have mentioned between its two proceſſes. The two bones of the fore arm are contiguous at the upper part, and after that are gradually ſeparated, and again meet together at the hand, changing there the proportions of their thickneſs, for the radius is fuller below, and the cubitus more ſlender. Laſtly the radius riſing into a cartilaginous head, is articulated to the end of the other. The cubitus is round at that extremity, but on one ſide has a proceſs that extends ſomewhat further. And not to be always repeating the ſame thing, we muſt take notice, that many bones end in cartilage, every joint is covered with it; by this proviſion they move upon a ſmooth ſurface, nor could they be joined together by ligaments or fleſhy parts, without the intervention of this middle ſubſtance.

In the hand, the firſt part of the palm conſiſts of many ſmall bones, the number of which is uncertain, but they are all oblong and triangular, and connected in ſuch a manner, that the angle of one, and the plain ſurface of another are alternately ſuperior, by which the whole together has the appearance of one bone, ſomewhat concave on the inſide, but two ſmall proceſſes from the hand are inſerted into the ſinus of the radius, and from the other ſide five ſtraight bones proceeding to the fingers complete the palm, from which bones the fingers ariſe, each of which conſiſts of three bones, and are all ſimilar one to another. The bones neareſt the palm are hollowed on the tops, and receive the ſmall tubercles of theſe more remote; they are all faſtened by ligaments, from which the hardened nails ariſe, having their roots fixed rather to the ſoft parts and not to the bone; ſuch is the nature and arrangement of the upper parts.

The lower part of the ſpine reſts upon the bones of the coxæ, which being tranſverſe and exceeding ſtrong, defend the womb, the bladder, and the ſtraight gut; they are gibbous on the external ſide, and hollow on the ſide next the ſpine; laterally, that is in the coxæ themſelves, they have round cavities, from which ariſes the bone called pecten, this ſtrengthens the tranſverſe part of the belly over the inteſtines at the pubis; it is ſtraighter in men, but more bent outward in women, to facilitate the birth. After theſe the thighs begin, the heads of which are more round even than thoſe of the arm bones, tho' theſe laſt are rounder than any other; below thoſe heads are two proceſſes, one before and another behind, then the bones become hard and full of marrow, and are gibbous externally; and they again ſwell into heads at the inferior part. The ſuperior heads are lodged in the ſinuſſes of the coxæ, as thoſe of the arms are in the bones of the ſcapulæ, thence they incline gently inwards towards each other, by which they ſupport the weight above in a more equal manner; and for the ſame reaſon the middle of the lower heads is hollowed, that they may be more eaſily ſupported by the bones of the legs. This joint is covered with a ſmall ſoft cartilaginous bone, called the patella, which hanging looſe and adhering to no bone, is fixed only by muſcles and ligaments, and inclining rather towards the thigh bone, defends the joint in all the flexions of the leg.

The leg consists of two bones, for in all things the thigh resembles the arm, and the leg the fore-arm, so that the proportions, and even the beauty of the one may be known from that of the other; which observation, as it begins in the bones, may be also extended to the soft parts. One of these bones is placed on the outside of the leg, and is properly called sura; it is shorter than the other, and slender above, but swells at the ancles. The other is placed forward, and is called tibia, it is longer and thick in the upper part, and this bone only is articulated with the inferior part of the femur, as the cubit is with the arm bone. The bones of the leg are also contiguous above and below, but seperated in the middle as in the fore arm.

The leg is received below by the transverse bone of the talus, which is placed above the heel bone; this in some parts is hollowed, and in others has processes, and receives the processes of the talus, and is mutually received into its sinus. The heel bone is hard, without marrow, and projects considerably backwards, and there approaches towards a round figure. The other bones of the foot are formed similar to these of the hand, the sole corresponds to the palm, the fingers to the toes, and the nails of the one to those of the other.

TO conclude the anatomy of Celsus, it is proper to add what we find relating to that subject in his general preface, where after having given a short history of medicine, he narrates in a most clear and beautiful manner, the opinions and disputes of the principal sects of the antient physicians, upon the foundations of medical knowledge, and among the rest of anatomy, of which the RATIONAL *or philosophical sect speak thus:*

BESIDES, as in the inward parts pains and various kinds of diseases arise, they are of opinion that no man can apply a remedy to parts he is ignorant of, and therefore that it is necessary to cut open dead bodies, and to examine their inward parts, and they extol Herophilus and Erasistratus who dissected criminals alive, given from prisons by the authority of kings, and while the breath yet remained, examined parts that nature had concealed; their situation, their colour, their figure, their size, their arrangement, their firmness, their softness, their smoothness, their processes and their cavities, their connexions, received by or receiving each other. Without this knowledge of the inward parts, how could any one distinguish what bowel was affected in any inward pain? and how could the cure be performed by one ignorant of the part affected? and if a man's bowels were exposed by a wound, how could the sound parts be otherwise distinguished from the injured ones, and the proper remedies applied, than by an exact knowledge of the natural colour of each part? Besides, by knowing the situation, the figure, and the size of the inward parts, external remedies can be more fitly applied; and the like reasons can be given for the other things that have been mentioned: nor can it be called cruelty, as some vainly suppose, by the sufferings of a few criminals, to find remedies for the deserving people of all ages.

IN opposition to the former, the EMPYRICS, *a sect equally respectable, who contended that experience in the practice of medicine was the only true foundation of the art, speak of anatomy in the following manner:*

NOW these things we have been talking of are only useless; but to open men alive is not only useless, but the greatest cruelty, and to pervert an art that has the glory of protecting the health of mankind,

mankind, to tormenting them, and that in the most terrible manner; especially when what is sought for with so much brutality, partly cannot be known at all, and partly may be learned without this barbarity: for the colour, the smoothness, the softness, the hardness, and all such things are not the same in the body thus cut open as it was in the entire man, because even without such violence, a thousand accidents, even of a smaller kind, make great changes upon the body; as fear, pain, hunger, crudity and lassitude, and it is much more probable that the inward parts are changed under such terrible wounds and butchery, as they are of a softer nature, and new even to the light; and is it not absurd to imagine that the state of parts is the same in life, as in a dying, yea even in a dead man? and allowing that while a man was yet breathing the abdomen could be opened, which is not the principal part, yet as soon as the knife advances towards the breast, and there cuts the transverse division (a membrane which divides the upper cavity of the trunk from the lower, the Greeks call it diaphragm) the man immediately expires, and so the butchering operator sees only the bowels and thorax of a dead man, because the parts must necessarily appear as in the state of death, not as when the man was alive; so the physician can only boast of cruelly murdering a man, not of knowing the state of the vicera during life: but if any thing useful can be seen while a man is alive, chance often puts that in our way in the course of practice; for sometimes a gladiator on the stage, a soldier in the field, or a traveller attacked by robbers, is wounded in such a manner, that in different men various inward parts appear, and show to a prudent physician, their situation, their position, their arrangement, their figure, and the like, not performing a murder, but a cure; and so he learns by humanity, what the others do by the utmost cruelty. And for the like reason, mangling dead bodies is by no means necessary, (which tho' it is not cruel, yet is loathsome) as most parts appear different after death, and what can be known from living bodies, we learn while we are curing them.

LASTLY, Celsus delivers his own opinion in a few words, agreeable to the sentiments of humanity, and of the greatest masters of the art:

AS these questions have been often and keenly handled by physicians in numerous volumes, and as the dispute still subsists, we must here subjoin what appears most probable, without partiality to either side, but taking a middle way, which is easy to be found, in this as in most disputes, by those that search after the truth in a fair and candid manner.——

NOW to return, I am of opinion that medicine should use reasoning, but should be founded upon evident causes; the obscure ones being removed, tho' not from the thoughts of the artist, yet from the practice of the art: but to cut open live bodies is both useless and cruel, tho' dissections of dead ones are necessary to learners; for they ought to know the position and arrangement of the parts, which dead bodies can better exhibit than living and wounded men; but the other things which can only be seen in living bodies, practice in the cure of wounds will discover, tho' more slowly, yet with more mildness and humanity.

NOTES

NOTES ON THE ANATOMY OF CELSUS.

BOOK IV. CHAP. I.

N. B. The pages and lines are cited as they are found in all the Latin editions of Celsus since that of Vander Linden in 1657.

Page 182, *line* 5. I HAVE added the titles of the chapters, tho' it evidently appears by many of them, that they are not from Celsus; besides, the most ancient manuscript of Celsus, of the seven in the Medicean library at *Florence*, neither has these titles nor any punctuation. There is no manuscript of Celsus that I know of in any British library.

Ibid. l. 14. I have taken the liberty to add here, at the end of the first paragraph, a few words taken from the beginning of the next chapter, as they show the intention and opinion of the author in this and the other anatomical parts of his work.

Ibid. l. 15—18. The Italian version, published at *Venice* in 1747, seems to favour the meaning I have given this passage.

Page 183, *l.* 1, *&c.* The comparing the rings of the aspera arteria to the vertebræ of the spine is natural and beautiful, tho' the resemblance is imperfect. These kind of comparisons are common with the ancients and with most fine writers, particularly Celsus, and have their use and beauty in science; thus the lungs are afterwards compared to an ox's hoof, an idea that has been retained by future anatomists.

As to the particulars of the anatomy of Celsus, contained in this chapter, besides the general beauty and elegance of the whole, I might mention several descriptions more just than those commonly received; for example, his concluding instead of beginning with the omentum, after having described the parts it covers, his considering the œsophagus, or gullet, as the beginning of the intestines, and his dividing the intestines into three, viz. the jejunum (commonly called the duodenum) the small, and the great intestines; a division which long ago occurred to me from nature and the simple view of the parts, when I was a very young anatomist, and before I had read Celsus, therefore I had more pleasure to find it in that admired author, who describes like a painter, as every true anatomist ought to do.

On the whole, how instructive and delightful is it to an anatomist, to see all the viscera of the thorax and abdomen thus presented to the eyes as it were in one view, and described in so short, clear, and natural a manner? I am persuaded that a good judge, who knows how difficult it is to describe in this masterly way, will be more pleased with this manner of Celsus, than with many tedious, unconnected, tho' laborious descriptions, which are too common in anatomy. The ancients were ignorant of the minute structure of animals, and of many small, tho' sometimes important parts, known to the moderns; nor were they accustomed to observe and describe with so much minuteness and accuracy, which indeed the moderns have carried to trifling and excess; but for judicious and elegant description, no modern can compare with the fine writers of antiquity. Education among them was complete and universal; eloquence was their peculiar study, and description is one part of eloquence, bringing things as it were before the eyes like the art of painting; and as the ancients did not know engraving, nor trusted so much to figures to supply the defects of their verbal descriptions, they were obliged to labour these to greater perfection.

BOOK VIII. CHAP. I.

Page 499. *l.* 3. and 4. *Offaque ejus, &c.* Thefe words muft either be tranflated in the fenfe I have given them, which I find is alfo the fenfe of the Italian verfion, fo as to underftand by *ab interioribus quibus inter fe connectuntur molliora funt*, the foft fpungy diploë, or internal fubftance that connects the two tables, or in the fenfe expreffed by the French verfion (Paris 1754), which no doubt the words will bear, but it is not confirmed by anatomy, nor is it even fo agreeable to the natural meaning of the words: thus, " *les os font plus durs à l'exterieur, & plus moux a l'interieur, vers les endroits ou ils s'uniffent. Entre les futures de ces differens os, s'infinuent plufieurs vaiffeaux, &c.*" As to the words HARD and SOFT, applied to bones by Celfus, tho' he feems once or twice to apply them improperly, which may be eafily excufed with other fmall errors, yet by HARD he in general means where the fibres are compacted and fmooth, and by SOFT where they are fpungy and form a cellular texture.

Ibid. l. 25—30. The words of thefe five lines are partly corrupted, as the fenfe of the latter part is obfcure; they feem alfo to be tranfpofed, being inferted in the middle of the defcription of the futures, to which they no way belong; for after *leniter infidunt*, *l.* 25. *at facies*, *l.* 31. very naturally follows; if we could therefore find the true place of thefe five lines, which feem really the words of Celfus, it might help to explain them: the moft likely place feems to me, *p.* 500. *l.* 30. after *audiendi eft*, where Celfus is defcribing the parts about the ear, and therefore naturally mentions the maftoïd procefs, and alfo the *os medium in exteriorem partem inclinatum.* I had once tranflated thefe words as a defcription of the temporal part of the zygomatic procefs, and this my conjecture of the tranfpofition of the words from where I have placed them, led me to; but as the whole zygoma is immediately afterwards defcribed, I rather adopted the opinion of Kraufe, which is ingenious, and not improbable, for we find Oribafius calls the fphenoïd, the middle bone, το μέσον ἀμφοτέρων ὀστοῦν, which bone, fays he, fome make to belong to the head, and others to the upper jaw, as it is fituated in the middle, μεταξὺ, between them; but tho' I have adopted the opinion of Kraufe, I by no means think my own without foundation; efpecially as I find the temporal bone, and its zygomatic and other proceffes, defcribed in the fame order in Oribafius, as here in Celfus; and yet immediately after, as in Celfus, is added alfo a full defcription of the zygoma. See the end of Oribafius's chapter on the bones of the head, and the fhort one that follows it on the zygoma. The lamdoïd future, tho' well known to the ancients, is omitted in Celfus, or more probably, is loft by the corruption of the text.

Page 502. *l. ult.* There is no need of putting *una* for *uno*, with Kraufe; but for *aliquid*, *page* 503. *l.* 2. I would read *alioqui*, with Linden, who fpoils this paffage by changing the ancient reading *promptum*, a word on feveral occafions elegantly ufed by Celfus, into *pronum*. This kind of liberty Linden has too often taken with Celfus, even in the cleareft paffages, a thing not to be excufed; by which, under the pretence of reftoring authors, they are mangled and corrupted in the groffeft manner.

Page 503. *l.* 34. *At a fumma cofta.* This paffage is furely much corrupted, and has given abundance of trouble to the editors and commentators upon Celfus, even to thofe moft converfant in anatomy. The later editions, I mean thofe copied from Vander Linden, by departing from the old editions after that editor, have corrupted many parts of this fine author, which have partly been reftored by Kraufe, merely by replacing the readings found in the old editions; but here Kraufe has retained the new paragraph of Linden, as if Celfus had been treating of a new fubject, and a new bone, whereas he is only continuing the defcription of the fcapula: and I am of opinion, that the words as they ftand in the firft edition of 1478, with a very fmall change, tho' no doubt fomewhat obfcure and corrupted, will appear to defcribe, firft the coracoïd procefs, and then the acromion of the fcapula, and in that fenfe I have tranflated them.

Page 504. *l.* 16. As in this part of the defcription of the *os humeri*, the editions differ greatly from each other, and from anatomy; I have chofen to follow Nature alone, as fhe appeared to me, in my tranflation of this paffage, having regard at the fame time to the manner of Celfus.

T Page

Page 504. *l.* 25. As *parvo excessu* is wanting in the first and other editions, I would expunge it, and for *extransitu* in the first edition, I would not read *extra situm*, or *extra id situm*, with Krause and Linden, but *extra in situ*, which approaches nearer the original word, and bears a much more simple, natural, and better sense, than to suppose that Celsus enters into so particular a description of the joint of the arm-bone with the scapula; and this is confirmed by his simple manner of describing the joint of the thigh. *Vertici lati*, for *verticillati* in one word, is an emendation that must be received by every one; and tho' obvious when discovered, by the ingenuity of the elegant Morgagni, was not thought of by any former editor; on the contrary, the spurious word *verticillati*, was not only explained and defended, but even inserted into dictionaries from this single corrupted passage.

GENERAL PREFACE OF CELSUS.

Page 7. *l.* 13. Much has been wrote and conjectured upon the word, *cortactum*, *contactum*, *contractum*, *confractum*, &c. all which words in this place are unworthy of Celsus: it would perhaps therefore be better to expunge the word altogether, as an interpollation and repetition, and to oppose *lævorem* to *processus deinde singulorum atque recessus*, which immediately follows it, especially as in *page* 11. *line* 13. where this very passage is as it were recapitulated, no such word is added after *lævorem*.

Had it been proper in so small a work, to expatiate on the beauties of this fine author, or to give the reasons and authorities that induced me to translate every passage as I have done, these notes, perhaps too tedious already, would have been still more so. I have no delight in verbal criticism, and nothing but a love of Celsus could have persuaded me to labour in studies of this kind. I shall therefore conclude these notes with observing, that as in this fine preface of Celsus, and in every other part of his work, we find the most masterly description of medicine, so we find in no part more striking proofs of his veneration for that art, and for those who were eminent therein; and also of his freely delivering his own sentiments on the most important questions in medicine: great arguments in favour of believing Celsus a physician, but a physician like a noble Roman, and in the most cultivated age of antiquity, with sentiments and ideas far above these that are too often found in the generality of the physicians of modern times, even in those who should be leaders and examples to the medical order.

IN Cicero's short sketch of the animal œconomy, we are not to expect the accuracy of a professed anatomist, much less the modern improvements; but we will find the true genius of philosophy, oratory, and sometimes almost of poetry. The moderns, in the whole of natural philosophy, are superior in matter and true doctrine, the ancients in the manner and art of writing, upon that and almost every subject. What rich and beautiful systems would they have left, had they been possessed of our materials! In my translation, I followed chiefly the edition of Lyons, printed in MDLIX; endeavouring, as much as I was able, to express the sense and beauty of the original.

CICERO OF THE NATURE OF THE GODS.

Book II. Page 128—138.

WE may clearly see, that the immortal gods have a peculiar regard to mankind, if we take a survey of his whole frame, of his outward form, and the perfection of his nature. The life of animals consists in three things, food, drink, and breath; the mouth is fitted to receive all these, and the nostrils to assist in breathing. The teeth, arranged in the mouth, chew, break, and grind the food; the sharp ones, opposite to each other in the middle, divide the morsel by biting it; but the inner ones (called genuini) grind it down; this seems also to be assisted by the tongue, at the roots of which lies the pharynx, or beginning of the œsophagus, here what we receive into our mouth first lands. This part is contiguous to the tonsils on each side, and is bounded by the extreme and inner parts of the palate, and by this is the food pushed down, being conveyed hither by the action and motions of the tongue, the parts of the œsophagus below the food being relaxed, and these above contracting themselves. But as the aspera arteria (for so it is called by physicians) has an entry contiguous to the roots of the tongue, a little higher than where the œsophagus is joined thereto, and as this passage reaches to the substance of the lungs, and receives that air which we breath, and sends it forth again from the lungs, this entry has a certain covering, to the end that no food may enter into this passage, and obstruct respiration. The abdominal bowels, immediately under the œsophagus, are the receptacle of the food and drink, but the lungs and heart attract the air which we breath. In the lower belly, there are many parts wonderfully constructed, and made up chiefly of membranes; it consists of many folds and windings, and holds and contains whatever it receives, either of moist or dry, so as easily to change and digest it; these parts are alternately relaxed and constricted, and they collect and intimately mix whatever they receive; so that by means of the considerable heat they are possessed of, and by grinding the food, and the force of the air, all is easily digested, melted down, and distributed over the whole body. The lungs are of a thin substance, and of a spungy softness, admirably fitted for sucking in air, they contract themselves in expiration, and are dilated by reception of air, that this animal food, the greatest support of living creatures, may be frequently drawn in. But the alimentary juice in the abdomen and intestines, being separated from the other parts of the food, is conveyed to the liver by certain direct passages, leading from the middle of the bowels to the portæ of the liver, (for so they are called) to which they reach and adhere, and from thence there are other passages reaching to the kidneys, by which the food passes that enters not the liver. After the bile and the liquors,

that

that come away by the kidneys, are thus ſecreted from the alimentary juice, the remaining part is converted into blood, and being collected at the ſame portæ of the liver, where all its paſſages meet, it paſſes into the vena cava, and is there blended; and being now prepared and digeſted, it is conveyed to the heart, from whence it is diſtributed over the whole body, by means of veins reaching to every part; nor is it difficult to explain in what manner the remains of the food are protruded by the contractions and relaxations of the inteſtines, but we omit theſe things that this diſcourſe may contain nothing unpleaſant. Let us rather explain that wondrous fabrick of nature, how the air that is drawn into the lungs by breathing, is firſt warmed by that very action, and then by the motion of the lungs; part of this air is ſent forth by expiration, part is received by a certain portion of the heart, which they call its ventricle, adjoining to that other which receives the blood that comes from the liver by the vena cava: and in this manner, from theſe parts, is the blood diffuſed thro' the whole body by the veins, and the air by the arteries; both which kind of veſſels, being many, frequent, and intimately woven over the whole body, ſhow the incredible excellence of this wonderful and divine work. Need I mention the bones, which as a frame hid under the other parts, are jointed in a wondrous manner, and fitted not only for the ſtability, and to determine the form of the limbs, but alſo for motion and every bodily action: add to theſe the ligaments by which the joints are bound together, and the muſcles interwoven and diſtributed over the whole body, in the ſame manner as the veins and arteries which come from the heart.

Much might be added of this attentive and ſkilful proviſion of nature, in order to ſhew what great and excellent things were given by GOD to mankind. Firſt by raiſing them from the ground ſtately and erect, that by contemplating the heavens above, they might attain a knowledge of the gods. For men are not upon the earth ſo much as mere inhabitants, but rather as ſpectators of things above and celeſtial, which are obſerved by no other animal. Then the ſenſes, meſſengers and interpreters of all things, are wonderfully formed for the neceſſary uſes, and placed in the head as in a citadel. The eyes like two centinels, obtain the higheſt place, from whence looking around they perform their office: and the ears for receiving ſounds, which naturally mount upwards, are rightly placed on high: ſo alſo the noſe for odours, which fly upward: and as the ſenſe of ſmelling is a principal judge of meats and drinks, it is rightly placed near the mouth; in which, as nature has opened a paſſage for the aliment, the ſeat of taſte is therefore placed. As to touch, it is equally diffuſed over the whole body, that we may be ſenſible of every impulſe, and every change of heat and cold. And as architects turn away from the view ſuch conduits as would be offenſive to the inhabitants, ſo nature has removed the like things far from the ſenſes. Now what artiſt but nature could have ſhown ſuch ſkill, in contriving the organs of ſenſe? firſt, the eyes are cloathed and ſurrounded with thin membranes, which are partly tranſparent, that we might ſee thro' them; and alſo firm, to ſuſtain the contained parts. The eyes are alſo ſlippery and moveable, that they might give way to ſhun danger, and be eaſily turned every way for the ſake of viſion; and that ſpot thro' which we ſee, called the pupil, is ſo ſmall, that it can eaſily ſhun things that might hurt it; and the eye-lids, which like curtains cover the eyes, are of the ſofteſt nature, ſo as not to hurt that tender organ, and moſt aptly made to ſhut, and prevent any thing from falling into the eyes, and alſo to open them, and this they can repeat with the utmoſt celerity; they are alſo fortified all round with hairs, as with a rampart; thus are the eyes defended when we are awake, and wrapt up under the ſame eye-lids, they are ſafely defended during ſleep. Beſides, the eyes are moſt uſefully ſunk, and defended all around with protuberant parts; above by the eye-brows, which carry off the ſweat falling from the head and brow, the cheeks gently

ſwelling,

swelling, protect them on the lower part and sides; and the nose is so placed as to resemble a wall between the two eyes. As to the ears, they are always open, for this sense is useful even in sleep, as we can be awakened by any noise. The passage to the ear is winding, for had it been straight, things might have more easily got in; and if any animalcule should attempt to enter, it is caught in the wax as in birdlime. What are commonly called the ears hang outward, not only for a defence to the organ, but also to catch the sounds, and convey them inward; they are formed of many hard and as it were horny cavities, with numerous windings, thereby increasing the sounds received, in the same manner as we add to the lute, shell or horn, and as from hollow cavities sounds are always increased. In the same manner the nostrils, always open for necessary uses, have a narrow entry to exclude noxious things, and they have a moisture useful to repel dust and the like. The sense of taste is well protected, being placed in the mouth, both for use and for safety. Now every one of the human senses greatly excel these of the other animals. First the eyes see every thing most exquisitly, in these arts of which they are the proper judges, as in painting, sculpture, modelling, and in the attitude and movement of bodies; the eyes likewise judge of the beauty and arrangement of colours and figures, and of their propriety and decency; they judge even of greater things, such as the virtues and the vices, they can distinguish a person angry or appeased, joyful or grieved, courageous or cowardly, rash or timid. The ears also judge in an admirable and skilful manner, by distinguishing the variety of sounds, both in the voice, and in wind and stringed instruments; the intervals, the distinctions, and every species of sound; the rich and strong, the dull, the smooth, the rough, the grave, the acute, the hard and the flexible; which differences are only perceivéd by the human ear. And we can judge of many things by the smell, by the taste, and by open touch; to please and satisfy which senses, there are too many arts invented; for we all see with regret, to what lengths unguents and cookery, and the luxury of the body have now arrived among us. Then with regard to the human mind and soul, he that does not clearly see a divine hand, in the reason, council, and forethought of man, must in my judgment be void of these qualities. While I am on this subject, I would wish, my friend Cotta, for your eloquence; how nobly would you expatiate here? first how great is our discernment, and then the power of reasoning, viz. joining and comprehending the consequences with what went before, by which we are able to conclude, with the greatest certainty, what follows from each thing, and are able to form definitions, comprehending them in a small compass; whereby we see the nature and the effect of the sciences, than which there is nothing more excellent, not even in the Deity. And how great are these things, which you academics endeavour to weaken and destroy, viz. our perceiving external things by the mind and senses, by collating and comparing which, we form all the arts for the use and pleasure of life. Then that mistress of all things, as you are wont to call it, the power of eloquence, how noble and divine! by which we learn things we are ignorant of, and teach others what we know. By this we exhort, we persuade, we console the afflicted, we banish fear, we check immoderate joy, desires, and anger; by this we were united in commonwealths, by rights and laws, and allured from a savage life. But without great attention, it is difficult to see the pains nature has taken in the organs of speech. First the windpipe reaches from the lungs to the bottom of the mouth, thro' this the voice passes, and is formed under the direction of the mind; then the tongue is situated in the mouth, bounded by the teeth, this stops and forms the profusion of the voice, producing clear and compressed sounds, by directing them against the teeth, or other parts of the mouth. Therefore orators say, the tongue resembles a plectrum, the teeth the strings, and the nose these horns which resound with the strings in music. Next with hands how fitly has nature provided us, and capable of so many arts! how easily can we stretch and bend the fingers, ſo that we perform every motion with eaſe, on account of the flexibility of the joints! therefore the hand and fingers are fitted for painting, for ſculpture, for muſical inſtruments of every kind. But what I have mentioned are chiefly for pleaſure, the hands ſerve alſo the neceſſary purpoſes of agriculture, building, manufactures of clothing both weaved and ſewed, and every work of braſs or iron: from all which we ſee, that by the invention of the mind, the perception of the ſenſes, and the labour of ſkilful hands, we are in poſſeſſion of every convenience. We erect cities, walls, habitations, temples; and alſo by the labour of man, that is, by the hands, we procure plenty and variety of food, for the earth bears many things acquired by the labour of the hands, to be immediately conſumed, or laid up and preſerved by art to prevent corruption; beſides, we feed upon animals, both of the land, the water, and the air, partly by catching, and partly by rearing them. And by training, we have the uſe of ſtrong quadrupeds, whoſe ſtrength and ſwiftneſs, we as it were convey to ourſelves; we put loads upon ſome animals, and yokes upon others; we turn to our own uſe the acute ſenſes of elephants, and the ſagacity of dogs; we dig iron from the caverns of the earth, a thing neceſſary for tilling the ground; we diſcover the hidden veins of copper, ſilver, and gold, metals both for uſe and for ornament. Then, by cutting trees, and wood of every kind, both wild and planted, we procure firing for warmth and culinary uſes, and alſo timber for building, to defend us from heat and cold; of great uſe alſo for ſhipping, by which we are ſupplied with profuſion of every earthly comfort; and the moſt violent things in nature, viz. the ſeas and the winds, man alone is able to govern, by his ſkill in naval affairs, whereby we enjoy many productions of the ſea. Man has likewiſe dominion over the products of the land; we enjoy the fields, we enjoy the mountains; the rivers, the lakes are ours; we ſow corn, we plant trees, we fertilize the earth by canals of water; we direct, we conduct, we avert the courſe of rivers; and by means of our hands, we as it were ſuperadd a new nature to nature herſelf. Yea, the reaſon of man has it not penetrated even to the heavens! for we alone of all animals, obſerve the riſing, ſetting, and courſe of the heavenly bodies; men have defined the day, the month, the year, the eclipſes of the ſun and moon; the times and quantities of which, they can predict to all futurity: from the contemplation whereof, men acquire a knowledge of the Gods; thence piety ariſes, to which is conjoined juſtice and the other virtues, and form all together a life of happineſs equal and ſimilar to that of the Gods, yielding in nothing to the cœleſtials, except immortality, without which life may be complete. From all which it is clear, how much mankind excel all the other animals, and that neither ſuch a figure and compoſition of bodily members, nor ſuch powers of ſoul and of genius, could ever have ariſen from chance alone.

FINIS.

DIRECTIONS TO THE BOOKBINDER.

Each Pair of Prints are to be placed facing each other, firſt the outline figure, and then the ſhaded one, in the following order:

The Pair marked TAB. I. between pages 22 and 23.
The Pair marked TAB. II. between pages 28——29.
The Pair marked TAB. III. between pages 32——33.
The Pair marked TAB. IV. between pages 38——39.
The Pair marked TAB. V. between pages 46——47.
The Pair marked TAB. VI. between pages 52——53.

The Binder is deſired to cut off as little as poſſible from the margin.

www.ingramcontent.com/pod-product-compliance
Lightning Source LLC
Chambersburg PA
CBHW021821190326
41518CB00007B/696
* 9 7 8 3 3 3 7 6 0 6 4 6 6 *